Handbook on Rabbit Production and Management

NIPA® GENX ELECTRONIC RESOURCES & SOLUTIONS P. LTD.
New Delhi-110 034

Handbook on
RABBIT PRODUCTION AND MANAGEMENT

Y. B. RAJESHWARI
MAHADEVAPPA D. GOURI
T.M. PRABHU

NIPA® GENX ELECTRONIC RESOURCES & SOLUTIONS P. LTD.
New Delhi-110 034

NIPA® GENX ELECTRONIC
RESOURCES & SOLUTIONS P. LTD.

101,103, Vikas Surya Plaza, CU Block
L.S.C. Market, Pitam Pura, New Delhi-110 034
Ph : +91 11 27341616, 27341717, 27341718
E-mail: newindiapublishingagency@gmail.com
web: www.nipabooks.com

For customer assistance, please contact
Phone: + 91-11-27 34 17 17
Fax: + 91-11- 27 34 16 16
E-Mail: feedbacks@nipabooks.com

ISBN: 978-81-19215-46-1

Composed and Designed by NIPA®.

KARNATAKA VETERINARY, ANIMAL & FISHERIES SCIENCES UNIVERSITY

Nandinagar, PB No. 6, Bidar-585401
Office: 08482-245264, Mobile: 09448133691
Fax: 08482-245241
E-mail: vckvafsu@yohoo.co.in
shkvafsu@yahoo.co.in

Prof. Suresh Honnappagol
Vice Chancellor

FOREWORD

I am very happy to go through the **"Hand book on Rabbit Production and Management"** Written by Dr. Y. B. Rajeshwari, Professor of Livestock Production and Management, Dr. Mahadevappa D. Gouri. Ph. D. Scholar, Department of Livestock Production and Management, KVAFSU, Veterianry College, Hebbal, Bangalore and Dr. T. M. Prabhu. Assistant Professor (SS), KVAFSU, Veterinary College, Hassan, Karnataka. In recent days lot of opportunities available for Veterinary graduates with sound knowledge of Rabbit management in different organisations especially those connected to Bio-medical Science Research and also the enterpreneure for both small and large scale rabbit farmers.

There was a long felt need to have up-to-date information in appropriate form. The authors have taken keen interest to fulfill the aforesaid necessity in the desired form in theis book. This book deals with different aspects of rabbit necessary elements like Breeds, Handling, Breeding, Feeding and Disease control. The information on the topics concerned is quite concise and comprehensive. The book also contains good charts, illustrations and figures. I hope it will be useful to veterinary and animal sciences students, teachers and scientists to enrich their knowledge. Appreciate efforts of authors in compiling this book.

It is my immense pleasure to recommend this book for general reading of the concerned field professionals in particular and others in general.

4th October 2010

Sincerely Yours

Suresh Honnappagol

PREFACE

It is our immense pleasure to come out with this edition of **"Hand book on Rabbit Production and Management"** The book has been authored in accordance with the syllabi prescribed by Veterinary Council of India (VCI), New Delhi for under graduate students of Veterinary Science and Animal Husbandry all over India. The book includes comprehensive and up-to-date information on the topics highlighted and also deals with animal ethics and welfare of Laboratory Rabbits and Pet Animals in precise manner and in simple language, which becomes easy for students to understand.

There are many books available on this subject covered in this text giving details of many aspects of topics, but definitely there was a long felt need of availability of updated information in concise form at single source. We have strived at our level best to fulfill this necessity of the subjects. Further, an attempt has been made to incorporate quality figures, charts, illustrations and photographs in the book as per need and availability. We hope, the text will be suitable and useful for students including those appearing for comprehensive examinations, teachers and scientists.

We humbly request our readers and users of this text to come forward by offering their suggestions and comments, which will be sincerely taken in positive spirit for improvement of next edition, if at all Almighty wishes to bring out the same in future.

Authors

CONTENTS

CHAPTER 1

Historical Background of Rabbitry

Rabbit domestication dates back no farther than the previous millennium. It were the Phoenicians who are said to have discovered the wild rabbit *Oryctolagus cuniculus* of Southern Europe and North Africa when they reached the shores of Spain about 1000 B.C. In Roman times, the rabbit was still emblematic of Spain. The Roman spread the rabbit throughout the empire as a game animal. Like the Spaniards, the Romans also ate foetuses or new born rabbits, which they called laurices. It then becomes a practice that the rabbits were kept in leporarian, stone -walled pens or parks for hunting purposes. These leporaria ultimately paved the way for the development of game parks in the middle Ages. The monks ate laurices during lent because they were an aquatic dish. In France, only the lord of the manor could maintain a warren.

Several breeds of rabbits were known in the sixteenth century and the fact is corroborated by records available that controlled breeding had been introduced by this time. As it was the most delectable dish of the monks, domestication of rabbits become a reality in the Middle Ages. It was in the sixteenth century that countries like France, England, and Italy also took to rabbit breeding. Experts had started classifying types of rabbits. The three types known at that time were

i) The wild rabbit

ii) The semi-wild or warren rabbit raised inside walls or ditches and

iii) The domesticated or hutch-bred rabbit.

However, it was only in the nineteenth century that rabbit rearing was undertaken by all and sundry and in almost all parts of Western Europe. It was now known in Australia and New Zealand also. In the earlier stage, rabbits were kept in the backyard, with the poultry.

In India, proven breeds of exotic rabbits were imported in fifties by private breeders to help the farmers to increase their income through rabbit farming. But the introduction of German Angora breed in early sixties brought a revolution in the popularization of rabbit farming in India, although it has been confined to cooler hilly areas mostly. It has generated tremendous interest among the farmers to adopt rabbit farming as a full time vocation or sometimes as a supplementary enterprise. The wool productivity was very low at that time but it picked up with the proliferation of rabbit farming. India produces around 80-100 tones of Angora wool per annum at present which is quite low still is meeting out the installed capacity of Angora spinning industry. Further requirement of wool is met out from the import from developed countries.

Surprisingly, rabbit farming has neither been included in any Five Year Plans nor there is any incentive to the farmers either at the State level or the National level. It appears that there has been no serious thinking as to the scope and expansion of Rabbitry. Nevertheless rabbit farming can open vistas of self-employment to the young educated and help the Government mitigate the plight of the marginal farmers as well. Rabbit excreta are the best source of nitrogen, phosphorus and potassium which can well reduce the burden of costly fertilizers.

As industry, rabbit farming is gaining popularity in cooler and temperate parts of the country such as Jammu and Kashmir, Himachal Pradesh and parts of Uttar Pradesh, and of late in northeastern states as well.

Special Features of Rabbits

- Rabbit farming is a minor agricultural, can be adapted in rural and urban areas as an enterprise.
- Since rabbits are highly docile and pleasing to look at, it attracts variety of people e. g. Small / Marginal farmers and urban breeders.
- Rabbits can utilize forage protein more efficiently compared to other livestock. They are efficient converter of feed or greens in to meat and wool than by other livestock.
- Rabbits have faster growth rates and better conversion efficiency with feed gain ratio as good as 1: 2-3 on dry matter basis. This is comparable with pigs and poultry.
- *Constant reproduction*: The rabbits can be bred any time of the year with peculiarity that a doe can be bred on the day of parturition. It can add 20-40 young ones / year, short gestation (30 days), faster weight gain and early sexual maturity are the three main factors. It has biological productivity of 10-25 times of its body weights in offspring's as compared to cow with 0.6 times and an ewe 0.8 times as body weight during one year period.

- *Meat characteristics*: There is no religious inhibition for the consumption of rabbit meat, which is pearly, highly delicious, comparatively low in fat and calories, but high in protein value. Young rabbit can attain about 2 Kg live weight in 12 - 14 weeks on high concentrate diet. The ratio of edible meat is about 50 % after slaughter. Back yard meat rabbit is nick named as "biological refrigeration" at farmers house. Nutritionally speaking this is one of the very best meats.
- The rearing of rabbits is comparatively cheaper to other livestock's. The cost of animal, hutches, sheds, feed etc, is comparatively less expensive.
- High degree of genetic variability in rabbits both within and between breeds. These differences are seen in growth rate, reproductive efficiency, disease resistance survival, maternal, qualities and productive traits (milk, meat, fur and wool). This wide variability in productive or reproductive traits is very much useful in breeding and selection programme. Short generation with higher prolificacy favors' quick genetic gain.

Constraints in Rabbit Production

- *Climate* : The rabbits lack functional sweat glands. Their thermoregulatory mechanism is also poor. Hence, at an ambient temperature of 100^0 F which is approximately close to their body temperature rabbit may die due to discomfort. Rabbit require between 5^0C to 32^0 C temperatures, with 60 to 80 per cent relative humidity. Thus heat being a major constrain for rabbit production. This can be very well controlled by insulated materials for house constrictions especially in large scale rabbitries.
- *Cages* : The construction of pucca houses requires investment of high capital in large scale rabbiteries. In small scale rabbits can be raised inside the buildings also or outside in hutches with a protective roof or cages made out of locally available bamboo splits or wire mesh.

- *Rabbit feed*: In small scale/backyard rabbit rearing, rabbit can be fed with vegetable kitchen wastes or fresh greens with little concentrates. But for large scale rabbitries have to depend on pellated feeding which can be expensive.
- *Marketing*: At present there is no well defined market for rabbits and rabbit products in country. Marketing of rabbit meat deserves due attention by the development agencies in order to popularize rabbit breeding in the country in big way.
- *Rabbit research/development*: For the planned and scientific rearing of domestic rabbit for meat, strengthening of research, development, extension and training programme have to be done by the concerned agencies in the country.
- *Disease susceptibility*: A major factor is that rabbits are susceptible to several diseases those reduce production to unprofitable level in many rabbit farms.

They are susceptible to respiratory diseases by Pasturella multocida, tuberculosis, psudo-tuberculosis, enteric diseases including enterotoxaemia and ceacal impaction. Viral diseases like myxomatosis, rabbit pox, parasitic diseases like Coccidiosis, tanaesis, mange are also common in rabbits. Thus there is huge morbidity and mortality in rabbits.

Decomposition of Rabbit meat : Blue – bottle or blowfly attack the flesh of rabbits and deposit ova. Thus maggots of appreciable size may develop from these fly in meat. Warm weather causes rapid decomposition of rabbit meat.

Breeding Aspects

Breeding age (Months) : Male: 6½ to 7, Female: 6 to 6½

Time to Mate: Early morning or evening, i.e. during cooler part of the day. The doe (female) should be taken to the buck (male) and not vice versa.

Mating : Natural, Forced, cage mating and by Artificial insemination.

Male: Female ratio : 1:10

Quantity of semen (m1) : 0.4 to 1.5 (AV.0.7)

Sperm production : The testes produce 5 to 250 million sperm each day.

Number of sperm per ejaculate : 10 to 300 million per ml. (AV.150 million).

Estrus cycle : Rabbits do not have a precise estrus cycle. (However they have a heat period).

Ovulation : Induced, 10 to 13 hours after successful mating.

Number of ova shed : Up to 20.

Pregnancy test : Palpating of abdomen between 10 and 14 days after mating.

Pseudo pregnancy : The doe will show symptoms of heat from 16 to 18 days after mating.

Gestation period (days) : 31 to 32.

Placing of nest box in doe's cage : On 26th day of Pregnancy.

Parturition time (Kindling) : Usually in the night or very early in the morning;

Number of kits (young ones) born per doe per kindling (litter size) : 2 to 12 (AV.8). Kits are born blind and hairless.

Kindling percentage: 50 to 75.

The kits open eyes on (day): 10 to 11.

Number of litters per year: 3 to 5 or 5.

Breedable economic life (years) of Buck and doe : 2 to 2½

Average Body Weight

- Birth : 60 to 70 g
- Weaning : 800 to 900 g

- 12 week old : 1.5 to 2.0 kg
- Adult Male : 4.0 to 5.0 kg
- Adult Female : 5.0 to 5.5 kg, Females are heavier than males.

CHAPTER 2

Rabbit as Pet

Brief History

Fondness of man for rabbits is well established by its appearance in the Aesop's fables and Vishnu Sharma's 'Panchtantra, both of which date back to 2nd century B.C. However, domestic rabbits are descendents of wild species of Western Europe and North Western Africa where *Oryctolagus* still exists. Interestingly, it was the Romans who first raised rabbits in leporaria (enclosed warrens) as a game animal. Like Spaniards, Romans also ate fetuses of newborn young ones whom they called laurices. These leporaria ultimately paved the way for development of game parks in the Middle Ages. The monks ate laurices during Lent because they were an aquatic dish.

Oryctolagus cuniculus is the only genus of domestic rabbit belonging to family Leporaria.

Hares and rabbits are commonly confused. In fact, a jack-rabbit is a hare and the Belgium hare is a rabbit.

Man domesticated rabbit for its multiple uses i.e. Production of wool, meat, fur and laboratory uses. It was only in the last century that it was used as a pet. It is a delicate, lovely creature that appeals to the finer sensibilities of human beings.

The rabbit are reared for the following purposes

1. For the production of highly nutritious meat for human beings.
2. For the production of good quality wool, fur and pelt.
3. For the use as laboratory animals for conducting basic and fundamental researches in biological, pharmaceutical, medical and veterinary services.
4. Rabbits can be used as toys and Novelties.
5. Its droppings can be used as fertilizer.
6. Rabbit can be used as a pet for companionship.

Terminology Used in Rabbit Husbandry

Rabbit = Any numbers of rabbit family irrespective of age, sex and breed.

Doe = An adult breedable female

Buck = An adult breedable male.

Litter = The total young ones delivered at a time.

Kit = A young rabbit whose eyes are not yet opened.

Bunny = A young rabbit below 20 weeks of age.

Broiler = Rabbit fattened to finish for meat at 10-12 weeks of age. These are fast growing breeds.

Fryer = 10-12 weeks old rabbit ready for market.

Rooster = Culled rabbit.

Kindling = Parturition.

Weaner = A newly weaned rabbit.

Fur = Wool.

Pelt = Skin.

The profitability in rabbit industry depends mostly on proper nutrition, breeding, housing, sanitation and disease control.

The Reason for Rabbit Keeping

Most of the countries in the tropics and sub-tropics need to produce more meat in order to supply their rising population with animal protein. The rabbit has several points in its favor as a meat producing animal.

Feeding

Rabbits can be fed on a great variety of locally available foods, weeds, some green leaves and kitchen vegetables can be given to rabbits at almost no cost. Commercial rabbitries and some rabbit keepers often buy more expensive concentrate rations for their rabbits.

Rapid rate of Production

Rabbits can breed and grow quickly. This means that a rabbitry can expand quickly and take advantage of plentiful food supply or an increase in the demand for rabbit meat. Even the backyard rabbit keeper can provide a constant supply of meat for his family.

Useful By-Products

Rabbit skins can be used for several purposes (mats, rugs and clothes). The manure is a very valuable organic fertilizer for use in the vegetable garden.

Herd Size

The number of rabbits in the herd can easily be matched to the rabbit keeper's resources. Even three does and one buck will provide a useful supply of meat for the family.

The Small Size of Rabbits

The small size means that they can be easily handled and cared for by women and children. It also means that the carcass is small enough to be used at once, without having to store any part of it under refrigeration.

Housing

House can be built from pieces of wood and bamboo. No special materials or equipment needs to be brought for the backyard rabbitry.

Acceptability : There are few religions or beliefs which do not allow the eating of rabbit meat.

The Disadvantages of Keeping Rabbits

There are very few disadvantages of rabbit keeping.

The Small Size of Rabbits

Means that they are more easily stolen or attracted by predators. They need to be protected from these by fences and other barriers.

Labour

Large numbers of rabbits require a lot of labour, particularly if they are kept in individual hutches. Each hutch has to be cleaned out daily. Each rabbit needs to be fed and watered two to three times daily.

Climate

Rabbits need protection against extremes of climate. They are especially sensitive to heat and sun, and always have access to shade.

Rabbit Production Systems

The exact way in which people keep rabbits differs from country to country, but there are three main systems of production.

The Backyard Rabbitry

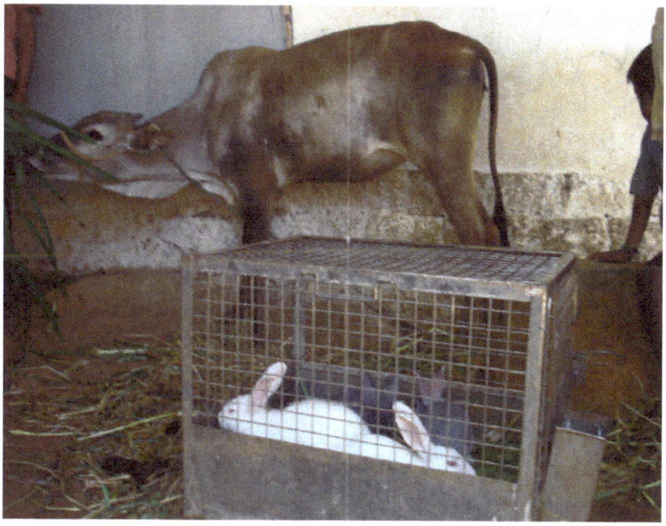

Is rabbit keeping on a small, family scale? A few does and one or two bucks may be kept in a home- built rabbitry just next to the house. The rabbits are fed on greens, weeds and vegetable kitchen scraps. Production from this type of rabbitry is low, but it provides enough meat to supplement the families' diet. It is a very useful way of keeping rabbits, especially in the poorer sections of the community, and provides a valuable supply of meat. Backyard rabbit production should be encouraged and improved. The incidence of disease can be reduced through better hygiene and production increase by better feeding and the introduction of improved strains of rabbit. Improvement is often limited by following constraints:

a) What can the small rabbit keeper afford to buy in the way of housing materials, feed and livestock?

b) What materials are available locally for low cost housing?

c) What cheap feeding stuffs are available for feeding the animals throughout the year?

As a result of improvements, some backyard rabbit keepers may be able to increase their productivity so that have some surplus rabbits which they can sell to their neighbours or at local markets. The cash received will be of help to raise their standard of living.

The Small Commercial Rabbitry

May have from ten to fifty breeding does in a purpose in a well built rabbitry. The aim of this type of rabbit production is to sell rabbit meat for a profit. The rabbits are usually fed on concentrates as well as bulky foods which cost money. In order to obtain a good profit, they must consider maters such as source and prices of rabbit feed, management techniques and markets. It would be unwise for someone to start up commercial rabbit farming, even on a small scale unless he had researched the project well and made good preparations. New rabbit farmers should be advised to start small with three does and a buck and to build up gradually. In such a way they gain valuable experience without taking too great a risk.

This type of rabbitry is more common in Europe and America. There are some examples of large rabbit units in the tropics, but to support such rabbits it is necessary to have a reliable market outlet for the carcasses, a source of good quality commercial feed, and expert veterinary services. As well as the meat they provide, these commercial rabbitries serve a useful purpose in the multiplication of breeding stock for distribution, the investigation of some of the problems concerning rabbit keeping in hot countries, and the training of local people in improved rabbit husbandry techniques.

The Development of Rabbit Keeping

Agricultural extension staffs are important people in the development of any programme aimed at increasing food production. This is particularly true in the case of rabbit farming because most town and rural dwellers know little about it and may find it hard to get information.

Scope of Rabbit Production

- Rabbit meat contains very little fat which is much desirable quality from health view point.
- Rabbit rearing is a fast return proposition.
- Wool, fibre, meat, fur and skin available from rabbit are used for preparation of garments, caps for children, ladies, toys, hand gloves, coat and dressing materials which fetch better price.
- Rabbits are very useful as laboratory animals for scientific biomedical research, embryo transfer to distant places / countries.
- Because of different size and colour like Havana, Flowering and English sport rabbit farming is now an attractive enterprise.

Rabbit Rearing

Rabbit production is a minor agricultural enterprise through out the world. It can be adopted in rural as well as in urban areas. The domestic rabbits are raised for meat, fur-skin and wool. In tropical countries it is primarily raised for meat. However, in our country, a different situation exists where rabbit production has not yet gained popularity. Firstly, because not many farmers know how important and profitable it could be as a meat animal and secondly, the domestic rabbits(white) are thought of as pets and only the wild ones(grey) for meat. Nevertheless, an interest has been generated among farmers of late to raise domestic rabbits. The potential of this species has apparently not been fully exploited and much remains to be done to enhance its role in small farm economy.

Rabbits are easy to raise and can be reared successfully by all strata of the society. They can adopt themselves to a variety of conditions, the capital investment needed is small when compared to most other livestock, the space requirements are little and inexpensive equipment can be used. The advantages of rabbits over other common livestock are their convenient size, high productivity and their suitability for back-yard production.

Rabbits are suited both to small-scale production and to large-scale intensive commercial production as well. However, in our present state of circumstances, rabbit production on relatively small-scale, involving minimal inputs could make a substantial contribution to the supply of animal protein for human consumption. Unlike beef and pork, rabbit meat is subject to hardly any religious restrictions.

The domestic rabbit is primarily herbivorous and can feed on a variety of grains, green forage, vegetable house-hold waste and agricultural by-products, unlike swine and poultry which need diets high in grain and low in roughage.

In our country scientific rearing of domestic rabbits is not yet well known. Consequently, the farmer is unaware of the rabbit's potential as a viable meat animal. To inculcate an interest and encourage rabbit raising, small production units on a backyard scale could be initiated to begin with. Small-scale (5 females and 1 male or 10 females and 2 males), low input system of rabbit production; using locally available materials seem to be relevant at present to the needs of our farmers and urban enthusiasts. When farmers and the general populace have become more familiar with rabbit meat, the foundations will be laid for a more soundly based commercial industry.

- Wild Rabbit - *Oryctolagus hispidus*
- Domestic Rabbit - *Oryctolagus cuniculus*

CHAPTER 3

Breeds, Breeding and Keeping Breeding Records

Breeds of Rabbit

There are many breeds of domestic rabbits in the world and all have different qualities. Out of the number of breeds and varieties about 38 breeds and 87 varieties are recognized and well established world wide.

For commercial purposes, the most important differences between rabbits are their size, colour breeding ability and suitability to the climate. Examples of several different breeds of rabbits are given below. Some of these might be used for improving local rabbit herds.

New Zealand White

This breed is commonly used in the production of meat. They are large rabbits, with meaty haunches and wide , deep shoulders. Adult buck may have weight 4-5 kg, and an adult doe weight 4.5-5.5 kg. They have white fur. This is the most popular breed in commercial rabbitries and is also widely used in tropical backyard units.

Californian

These are fairly large rabbits used often in the production of meat. They are well fleshed on the back and haunches and usually have a better meat to bone ratio than the New Zealand White. An adult Californian rabbit weight 3.6-4.8 kg. This breed has a white body with black on the nose, ears, feet and tail.

Giant Blanc

These are very large white rabbits and are sometimes match with Californian rabbits and New Zealand White rabbits to produce a large hybrid which is good for meat production. An adult Giant Blanc buck weight 6.3-7.3 kg and an adult doe weight 6.8-7.8 kg. They have white fur.

Flemish Giant

These are also very large rabbits which are reared for meat production. An adult Flemish Giant has a very wide back and weight 5.3 kg.

Beveren

These rabbits are large, with developed meaty joints and black haunches. They are useful in meat production. An adult Beveren weighs 3-3.6 kg.

Chinchilla Giganta

These rabbits are fairly large and light grey in colour. They mature early and have a good meat to bone ratio. An adult Chinchilla weighs about 5 kg.

Argente Champagne

These are medium sized rabbits weighing about 4 kg. They are kept mainly for their fur which is light or dark grey.

Dutch

The Dutch rabbit is a small, compact, early maturing rabbit. It has very good mothering qualifications and has been used in meat production for this reason. An adult Dutch rabbit weight about 2.3 kg. They have a half black and half white coat.

Havana

These are also fairly small, compact rabbits, but they produce well developed, meaty joints. An adult Havana weighs 2.7- 3 kg.

New Zealand Red

These rabbits are of medium size, with good mothering qualities and a good meat to bone ratio. An adult rabbit weights 2.7-3.6 kg. They have reddish brown fur.

Despite the good results shown by broiler rabbit in adaptation, production and reproduction traits it has not created a good impact on the farmers of North India. In southern states like Karnataka, Andhra Pradesh, Tamil Nadu and some north eastern parts of the country, the rabbits for production are getting popularized and adapted scale can easily be adapted by small / marginal farmers, urban poor and tribal's thus improving their socio-economic condition of this section of society, whereas, big organized commercial rabbitries can come up around big cities of the country where consumption, marketing and preservation of white meat may not be a problem.

Angora

This breed of rabbit is kept entirely for its wool, is usually white. They are small rabbits, weighing around 2.5-3 kg. Thc average wool production from this breed has gone up to 700-800 g per animal per year in India. There are different varieties like Russian Angora, English Angora and German Angora.

English Spot

New Zealand White

Russian grey giant

Chinchilla

Havana

Angora

Figure 1. Different Breeds of Rabbits

Breeding of Rabbits

(Mating, Pregnancy and Birth)

Introduction

For a rabbitry to be as productive as possible, it is important to systematically manage reproduction.

Many does do not produce as many litters as they could because they are mated at the wrong time. Many young rabbits die because they are not cared for properly. The farmer can overcome these problems by giving attention to the reproduction of his or her rabbits.

In this chapter we will look at

- The Buck
- The Doe
- Mating: when and where this should take place.
- Pregnancy and false Pregnancy.
- Birth.
- Lactation.
- Feeding of doe and her litter.
- Fostering.
- Failure to breed.

(a) The Buck

The buck is the male rabbit. A buck is old enough to use for breeding when he is eight months old, and should continue to be a good breeder for at least two to three years. A young buck 8 to 12 months old, should only mate one doe every third day. A mature buck from 12 months old can successfully mate four to six does a week.

If a buck is working as hard as this, he may start to lose weight, so give him more food. It is important not to let a breeding buck become fat. Never keep two adult bucks together because they may fight.

The buck should be selected on seeing the breeding records and pedigree. The selected buck should have two well developed testicles. He should be in good physical condition, good health and growth rate.

Breeding Systems

After a careful selection of breeding animals, they should be bred according to definite breeding plan. Various breeding systems used in rabbit to production are discussed below.

1. *Inbreeding*: Mating of related animals to produce uniform inbred lines. The detrimental effects of inbreeding are higher rate of mortality, low reproductive efficiency and lower meat/wool production.
2. *Line breeding*: It is also a sort of inbreeding and used for producing purebred lines. It leads to increased genetic uniformity with lesser chances of undesirable traits. The general objective of the line breeding is to maintain the line related to particular animal having the most desirable traits.
3. *Out breeding*: It is a system of mating/crossing of unrelated lines within the same breed. This is practiced when rabbits in a line start developing faults in productive/reproductive traits, out breeding with an other line traits tend to correct deleterious characters.
4. *Cross breeding*: This system involves mating of animals of different breeds to take advantage of heterosis and nicking ability. The rabbits of inferior breed are mated with superior breed to improve productive/reproductive traits.
5. *Rotational crossing*: In this system of breeding several breeds are used for crossing in rotation to bring about desired amount of heterosis and nicking ability.

(b) The Doe

The doe is the female rabbit. After mating with the buck she will usually become pregnant. 29-32 days later she should give birth to a litter. A doe selected for breeding should have eight functioning teats, so that she can feed eight kits easily. She should be in good physical condition and health. She should not be fatty. The best time to select new breeding does is when they are 6-7 months old and weigh about 2.5-3 kg.

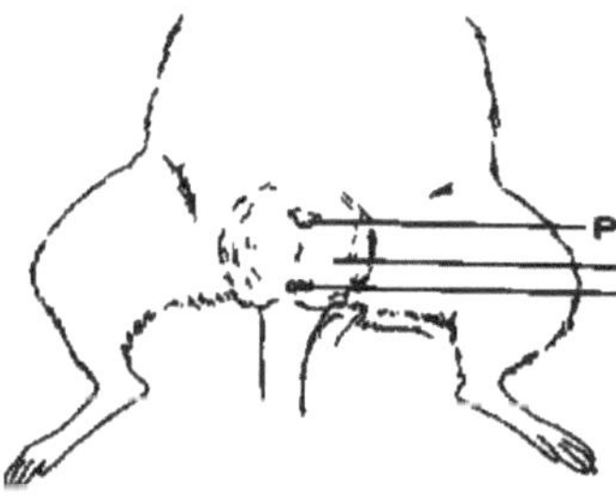

Figure 2. External genitalia of the male gerbil. P- tip of the penis, S-Scrotal sac, A- anus.

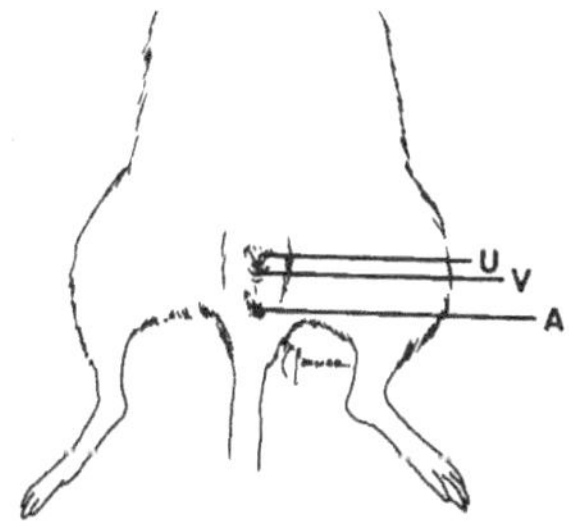

Figure 3. External genitalia of the female gerbil. U- Urethral Orifice, V- Vaginal Orifice, A- anus.

(c) Mating

Frequency of heat: Does will only become pregnant if they are mated while they are on heat. Adult does are on heat for 12 days then they have four days, when they are not on heat, after this they are on heat again, forming a 16-day cycle.

Signs of heat: It is important for a farmer to recognize the signs of heat in a doe so that he can take the doe to the buck for mating at the correct time.

The signs of heat in a doe are:

i. The doe may be restless.
ii. The vulva may look red and swollen.
iii. The doe will allow the buck to mate her.

Age at First Mating

Young does will come to heat for the first time at about six months of age, but they should not be mated until they are more fully grown(mature). After weaning the young male and female rabbits must be kept separate to prevent mating.

The age at which rabbits are mature enough to be used for breeding depends on:

i. *Breed*: small breeds mature earlier than large breeds.

ii. *Nutrition*: well fed rabbits mature earlier than poorly fed rabbits.

The following table may be used as a general guide

Breed	Approximate weight when ready for mating	Approximate Age when ready for mating
Small	2.5 kg	4-6 months
Medium	4 kg	6-9 months
Large	5 kg	9-12 months

Mating : Always take the doe to the buck for mating, because if a buck is put in the doe's cage, she may fight him to protect her cage. Early morning and the cool of the evening are the best for mating.

If the doe is on heat, she will raise her tail and let the buck mate her. After mating, he will fall off.

The buck and doe should be watched to make sure that mating takes place , and the doe then returned to her hutch. They should not be left together, even if mating does not occur at first. It is better to return the doe to her hutch and to bring her back again to the buck later.

If the doe refuses to let the buck mate her, it is possible that she is not on heat. Wait three to four days , then take her to the buck again.

If the doe persistently refuse to allow the buck to mount her, she can be restrained. The ears , and a fold of loose skin over the shoulders , are held with one hand and the other hand passes under the doe's body and between her legs. This hand supports her weight and raises her rear quarters to a convenient height for the buck. By placing the thumb and forefinger on either side of the vulva and pushing the skin gently backwards, the doe is induced to raise her tail which allows the buck to mate. Fig. shows how the doe should appear to the buck in this position. One successful mating is all that is needed but some people like the buck to mate the doe a second time about two hours later. This is not really necessary. After mating , fill in the record cards so that date when she is due to give birth is known.

Frequency of Mating

i. *Doe :* This depends on whether the rabbitry is commercial or backyard.

   a. *The backyard rabbitry*: The farmer should be aim for four litters per doe per year. He can do this by weaning each litter at seven weeks old. Mate the doe during the week after weaning. The complete breeding cycle (from one mating to the next) would take about 80 to 87 days.

      If the doe becomes pregnant at each mating, it would be possible to get 4 litters from her in a year. However, the actual number of litters may be less than this because of unsuccessful matings.

   b. *The commercial rabbitry*: These farmers usually aim for five or six litters per doe per year. This can be achieved in two ways:

      1. By weaning the litter at five weeks old, and mating the doe in the week after weaning. The doe can produce almost five litters per year this way. Each breeding cycle would take 66 to 73 days.
      2. By mating the doe 21 days after kindling. She will still be suckling her litter when she is mated. The litter should be weaned about two weeks after the

mating so that the doe can use all her energy for pregnancy. The litter would therefore be weaned at about five weeks old. In this way the doe can produce about seven litters per year. Each breeding cycle would take 52 to 59 days.

However, this method is very hard work for the does. It should only be used by rabbit keepers with very high standards of feeding and management, otherwise the does get and become ill.

Ii is suggested that the small commercial rabbit keeper should aim for five or six litters per doe per year.

If a litter dies, the doe can be mated again in four or five days. The dead baby rabbits and their bedding should be burned and the hutch thoroughly cleaned with disinfectant.

Examine the doe carefully. If she is ill at all , treat her and wait until she is completely better before mating her again.

If a doe loses the next litter as well, and reason cannot be found, she should be culled.

ii. Bucks should not be over-used; neither should they be under-used. Either could result in infertility. A young buck should perform one or two matings per week. A mature buck may be used up to six times a week, but not more than twice in the same day.

(d) Pregnancy and False Pregnancy

Pregnancy

Duration of the pregnancy - the length of pregnancy in rabbits is from 29-32 days.

Signs of pregnancy : It is helpful to know for sure whether a doe which has mated has become pregnant or not.

One method of doing this is to test mate the doe. This is not very reliable because some does will accept the buck even when they are pregnant while others will not mate again even if they are not pregnant.

A more reliable method is known as palpation in which the developing embryos (baby rabbits) can actually be felt by hand. This is a job for an experienced rabbit keeper because the delicate embryos can be easily damaged, if handled roughly.

Palpating the doe can be carried out in her hutch or on a table top covered with sacking to prevent the doe slipping. It is best done about two weeks after mating when the embryos should be the size of large beads (about 1-2 cm diameter), if the doe is pregnant. The embryos can be felt by exerting gentle pressure with the thumb and forefinger on either side of the uterus. The technique depends on some knowledge of the doe's anatomy together with considerable practice. During the last two weeks of pregnancy the doe's belly will become longer and her teats will gradually swell.

Care of the Pregnant Doe

Immediately after mating the doe should be returned to her hutch. If her previous litter has not yet been weaned. She can go back to them and continue to suckle them until a week before her next litter is due. If the doe is on her own during her pregnancy, she can be kept in a smaller single-size hutch for the first three weeks. At least a week before kindling is due , the doe should be in a breeding hutch to get used to the place where she will give birth. At the same time a nest box and suitable soft materials for marking a nest, such as grass, hay or straw should be provided. It is important not to use sawdust as it will get into the young rabbits eyes. Wood shavings, however, are excellent. A careful watch should be kept on the doe to see that she starts to make a nest.

The day before kindling the doe will pull out some of her fur from her rump, sides and around her teats. She will use this to add to the nest.

Feeding of Pregnant Doe

Feeding should be gradually increased during the last 10-14 days of pregnancy because the growth of the embryos and the development of the doe's mammary glands will be taking place rapidly during this period. Good feeding at this time is likely to increase the birth weight of the young rabbits and milk yield of the mother.

False Pregnancy

If a doe did not get pregnant when she was mated, she may develop a false pregnancy. This may also result if another attempted to mate her. She will refuse to mate. After 16 to 20 days of mating, she will start to pull out her fur and try to make nest. (A pregnant doe usually starts to pull out her fur about 25-28 days after mating.) Soon after this , she will come to heat again and can be mated.

Aggression Behaviour by Does

Although this may cause handling difficulties, it is usually a sign of a good mothering ability.

(e) Birth (Kindling)

Signs that the time of birth is near: The rabbit keeper will know the date when the doe is due to kindle from his records. The day before kindling the doe will pull from her rump, chest, fur/wool sides and teats to prepare the copy nest for kits.

Kindling

The birth of the young rabbits is known as 'kindling or parturition.' It often takes place at night. There are usually no problems, provided the doe has the materials and place to make a nest, and is left undisturbed. It is particularly important to ensure that predators are kept away at this time because, if the doe is upset or frightened, she may easily kill or abandon her young.

As each baby rabbit is born, the mother licks it and allows it to suckle. The babies are born naked and blind and weigh only 30-80 gm , depending on the size of the mother, how she has been fed during pregnancy, and on the number of babies in the litter. The number can vary between two and twelve.

A doe cannot successfully rear more than eight kits because she only has eight teats. Select the strongest eight young rabbits from the litter and leave them with the doe. Take the others away and either foster them or cull them. The lactating doe nourishes the young from eight teats once in the morning and that too for 5-10 minutes only.

The doe's use of the nest box is very important. Sometimes doe may not make a nest at all and simply gives birth on the floor of the hutch. This may be due to some disturbance or to the fact that she is a bad mother. The babies are likely to die from cold or neglect. If they are placed in the nest box, with some soft material, there is a chance that the doe will take care. If not, the only alternative is fostering.

DoeZs usually kindle or parturate without any problems. An experienced rabbit keeper should be called if there are any problems. Sometimes a doe will kill her young ones (cannibalism) or abandon them for no reason. If she has not done it before, mate her again in four or five days. If this should happen twice with the same doe, she can be classed as a bad mother and should be culled from the herd because this is a characteristic that can be passed on to her progeny. Balanced feeding adequate watering and cleaning will reduce stress on mother, this can prevent such incidents.

Care of the Young Rabbits

The litter should be inspected every day to make sure that the baby rabbits are well fed and comfortable. If the babies are hungry or too hot, they may be restless and start trying to climb out of the nest box. If they are well fed and comfortable they will be sleeping quietly.

After 10-12 days, the baby rabbits eyes usually open, and their fur starts to grow.

At about 21 days, the young rabbits will start to come out of the nest box.

(f) Lactation

The mammary glands of the doe develop during pregnancy particularly during the last week. They start to secrete milk just before the birth of the young.

Milk is the only food for the young rabbits during the first two or three weeks of their lives. Their development during this important period depends on the mother's milk supply. This is influenced by characteristics passed on from the doe's parents and type of feeding management during the last half of pregnancy and during lactation.

(g) Feeding the Doe and Her Litter

Food

Good quality feed should be provided *ad libitum* during rearing period. For the first three weeks this will be eaten entirely by the doe. During this time the only food for the baby rabbits is the doe's milk so it is important that she receives enough quantity of good food to enable her to secrete milk for her babies. After two to three weeks, the baby rabbits will start to eat the grass and concentrates, as well as suckling the doe. As they get older, they will eat more grass and concentrates, and suckle less milk from the doe, until they are weaned.

It is a great advantage if concentrates can be provided during the whole of the rearing period. The mother will give more milk and the babies will grow faster than would, if grass were the only food available.

Water : A plentiful supply of fresh, clean water is most important for the suckling doe and her litter. In order to produce plenty of milk the doe must get enough water to drink. A shortage of water will also reduce the amount of feed she is able to eat which will reduce her milk production. In hot weather a doe and her litter may drink up to 4.5 liters of water in a day.

(h) Fostering

This is the transfer of baby rabbits to another doe for rearing. It may be attempted for any of the following reasons:

i. The mother dies, or abandons the young, soon after birth. The young rabbits would die from starvation unless a foster mother could be found.

ii. The litter is too large (more than 8) for the mother to feed all the babies. The extra babies can be transferred to another smaller litter.

iii. Two, or more, does may kindle at about the same time, having litters of different sizes. Distributing the babies more evenly between the mothers would probably result in better overall growth of the young rabbits.

Fostering does not succeed in every case, but if the following method is used, there is a good chance of success.

a. Rabbits to be fostered should be less than one week old, though fostering is occasionally successful at a slightly later stage.
b. The age of the baby rabbits to be fostered should be nearly the same as the foster mother's own litter(not more than three days difference in age).
c. Fostering is carried out by moving the young rabbits to the foster mother's nest box. She should not be in the hutch at the time. Rub these young rabbits with the nesting material in the foster mother's nest box, to make them smell the same as her own young. Leave them for several hours, then put the doe back in the hutch.
d. Darken the hutch and leave undisturbed until the following day.

If all attempts at fostering fail, it is better to destroy these extra rabbits than to let them suffer, because they will die without a mother anyway.

(i) Failure to Breed

Occasionally a doe will not become pregnant after mating. There are many causes. The most important are:

Age : The buck or the doe may be too young. A guide to the correct age for first mating, also, the buck or the doe may be too old, so it should be culled.

On heat : The doe may not be on heat. Try mating her again in about 4 days.

False pregnancy: If the doe starts to pull her fur out 16 to 20 days after mating, she has probably developed a false pregnancy.

Physical condition : A rabbit which is too fat or too thin is less likely to breed than a normal rabbit. Look closely at the feeding; it may need to be changed.

Disease or injury : A rabbit showing any signs of disease or injury should not be used for breeding. It should be treated or culled, depending on the condition. It should only be used for breeding again when it has completely recovered.

Moulting : Rabbits which are moulting are less likely to breed successfully. Try again when they have finished moulting.

Sterility

Occasionally a buck or doe may be permanently completely sterile. The farmer can use the records to see which rabbits are sterile.

If a buck has failed to make several does pregnant, he is probably sterile. If a doe has failed to become pregnant after mating with several different bucks, she is probably sterile.

In both cases an experienced rabbit keeper or a veterinarian should be called to examine the rabbit. If it cannot be treated, it should be culled.

In general, if a doe is not pregnant after mating, mate her again with a different, more vigorous buck.

Hot Weather

During very hot weather when temperature reaches above 30°c, especially when the relative humidity is also high, breeding can be impaired. These conditions cause stress and can cause a temporary sterility in male rabbits.

CHAPTER 4

Care of Young Rabbits

Introduction: Most young rabbits will be used for meat and skins; some will be kept for breeding. Whatever they are used for, the rabbit keeper will benefit if they are well cared for.

Handling of Rabbits: There will be times in any rabbitry when the animals will have to be picked up, e.g. when a doe is moved to a breeding hutch or when young rabbits are weaned.

When it is necessary to pick up a rabbit for any purpose, the aim should be to cause it no discomfort and to prevent it struggling. So it must be handled gently but firmly. There are three ways to hold a rabbit according to its size.

Young Rabbits: Up to the age of about eight weeks young rabbits may be picked up and carried in one hand by gently grasping the loin. Rough or careless handling by method can cause serious damaged to the rabbits. Do not pick up any rabbit over eight weeks old this way.

Two-hand grip for restraining a rabbit. Rabbits improperly restrained will struggle and may luxate or fracture the lumbar spine.

Medium-sized rabbits: The rabbit may be caught by grasping a loose fold of skin over its shoulders. Then place the other hand under the rump to support it before lifting.

Large Rabbits: Rabbits weighing over 5 kg may be caught in the same manner as above but the other arm should be extended along the side of its body with the hand under the rump for support. The attendant should hold the rabbit against his own body when lifting, to provide extra support.

On one account should any rabbit be lifted by its ears because it is very painful for the rabbit and can severely damage it? The rabbit should always be held facing away from the handler as it has sharp claws, and a kick from the hind legs can cause a bad scratch.

Feeding Young Rabbits: The feeding of adult rabbits is discussed in the next chapter. In this section , we will look at the special needs of young rabbits.

Suckling: For the first two to three weeks, the baby rabbit's only food is their mother's milk. It is important that each baby rabbit can suckle and get enough milk. If the doe has too many babies or not enough teats, some of the babies will be hungry and may die. The doe must be given plenty of food and water so that she has enough milk for her litter.

Solid food: Form two to three weeks old, the young rabbits will start to eat grass and concentrates as well as suckling the doe. As they get older, they will eat more solid food and suckle less from the doe. Plenty of fresh food should be available for them two to three weeks old.

Weaning: At five to seven weeks old , the litter is separated from the doe. Usually the litter is taken away leaving the doe in the hutch. Some times farmers prefer to leave the litter in the hutch and remove the doe so that weaning is less of a shock to the litter.

From weaning onwards the young rabbits can be fed green food, vegetables and concentrates.

Routine Attention to Young Rabbits : At the same time as weaning several other tasks can be done.

Determining the sex of young rabbits: This is as 'sexing' and is usually carried out at the time of weaning when the sexes are fairly easy to distinguish. The young rabbit is balanced on the forearm with the hand under the rump. The thumb and forefinger of the other hand press down gently on either side of the sex organ. In bucks the penis will protrude as a rounded tip, whilst in does the protruding vulva will appear as a slit.

Sexing much younger rabbits can be done but it is more difficult and there is seldom any need for it.

Weaning Young Rabbits

The purpose of this is to obtain a record of how well the mother has reared her litter. Such records are very useful later in order to compare the breeding does and decide which are the best mothers from which replacements should be kept. The weaners can all be weighed together in a basket, or small cage.

Tattooing: In order to be able to identify young rabbits after weaning, it is necessary to mark them in some way. The best method is to tattoo a number in ink inside the ear of the rabbit. This can be done by writing the number with a special needle, or by using the equipment.

Castration: If young male rabbits are kept in the same cage after weaning, they are likely to start fighting. There are two ways of preventing this separation or castration. In the case of rabbits kept for meat, they may reach slaughter weight before any out break of fighting occurs. If not, separating the rabbits is much better than castrating them.

Castration is only necessary if male rabbits are to be kept together for a long time such as in the case of Angora rabbits kept for the production of wool. It is not an easy task, and should only be done by an experienced rabbit keeper or veterinarian.

Breeding does: The young does which the farmer has selected for breeding can reared as a group in a large hutch. One month before the date of the first hutches. This will prevent them having false pregnancy. These young does should be fed well so that they grow quickly, but do not let them get fat.

Breeding bucks: Most of the young bucks will be used for meat and skins. Some may be selected for breeding. These bucks should always be kept single hutches to prevent them from fighting. They should be well fed that they grow quickly, but do not let them get fat.

Rabbits for Meat Production: Rabbits which are kept for meat production can be kept in groups in large hutches. Males and females should be kept apart to avoid fighting and pregnancies. Several does can kept together in a large hutch until the time of slaughter.

Sometimes several young bucks can be kept together in a large hutch until about three old. After this age, they will often start fighting. This can stopped by separating the bucks and keeping them bit keepers castrate young bucks, because it is difficult procedure and because bucks which are not castrated grow quicker and produce better carcasses for meat.

If the bucks are slaughtering at two to three months old, the problem, of fighting is not likely to arise. Rabbits kept for meat production should be fed well so that they grow quickly, but do not let them get fat.

CHAPTER 5

Systems of Housing for Rabbits

A suitable site for housing is very important. It is better to select a site which has good drainage, adequate shade like under trees, and with sufficient surrounding space from strong winds. Basically the shelter must protect rabbits from extreme temperatures and predators like snakes, mongoose, dogs and other animals.

Rabbits can be raised inside the building also or outside in hutches with a protective roof. A hutch is a small building containing cages for housing a few rabbits (6 to 12) a back-yard scale. They are generally made of wood or stone one and chicken wire mesh. The floor may be made of a stronger ranger mesh or soil floor bedded with straw. Hutches can also be constructed with local building materials such as tree and woven split bamboo, etc. The bamboo and wood cages should be 48 inches long, 30 inches wide and 24 inches high for each adult rabbit. These cages should stand on upright supports sports waist high. Mats or gunny bags could be placed if hanging the sides of the hutch to protect the rabbits from sun and rain but not block ventilation. The roof should be sloping and could be made of Palmyra leaves or grass, good ventilation is necessary to prevent respiration and health problems in the rabbits.

Wire - mesh floor has the advantage over solid floor bedded with straw in that the animals waste drops down through mesh openings and thus keeps the cage dry, clean, free of ammonia odour and fly menace. The leg supports for cages should be treated with tar and the wood used for cage treated with old used motor oil to make the hutches resist insects and weather changes.

Rabbits can also housed in that roofed mud huts or hard mud floors, with a small yard and mud floors, with a small yard and adequate ventilation. It is also possible to construct rabbit housing from materials such as old packaging cases, intermeshed branches bamboo strips or any locally available hard wood.

In colony rearing, several rabbits (10 or more) could be kept together without confining them in cages. A floor space of 1.5 to 2 square feet per rabbit is sufficient. Rabbits could be reared on wire-mesh floor or on a hard mud floor in a deep litter fashion. Supervision is easier with such a system and is comparatively less expensive. In colony rearing all the females can be housed together. However, for pregnant mothers, from the time of kindling until weaning bunnies, it is advisable to keep her and the litter isolated from other adults. All males when housed together often flight amongst themselves. So, colony rearing is not recommended for adult males. Weaned and growing rabbits of either sex meant for sale can be housed together safely. The main disadvantage of colony rearing especially if it is the litter type is the greater risk of coccidiosis and maintenance of cleanliness.

In western countries, hutches and cages are made of wire mesh. They are constructed either in a single decked, double or 3 decked (tier) fashions. Two tier or three cages, though expensive, are space savings. Such arrangements are in vogue in material rabbitries. The material wide used is the galvanized welded mesh. The wire-mesh ½ inch of 16 or 14 gauge is used for floor very particularly. This particular mesh prevents the feet of the rabbit getting entangled in the mesh and at the same time allows the faeces and urine of the animal to drop down easily. The common arrangement in a rabbitary consists of rows of cages in a well lighted and ventilated shed with two rows opening into a central walkway which makes feeding and watering easier.

The dimensions of the rabbit wire cage for a doe of medium breed like New Zealand white should be 36 x 30 x 20 inches. For the sides and top of the cage, welded wire-mesh of 1 x 2 inches and for the floor galvanised welded wire-mesh of ½ x 1 inch or ½ x ½ inch is used. The gauge should be 16 minimum. The door of the cage should be on top and be 18 x 15 inches to allow access to all parts of cage inside and for introduction of nest box.

Bucks should always be kept separately in 30 x 30 inches cages. Some farmers use a circular round buck, so that during mating, the doe can't get into a corner to prevent the buck from breeding.

Young weaned rabbits during the growing period should be housed in groups of 5 to 6 per square yard of cage floor, besides , it is a good practice to have a couple of isolation cages 30 x 30 x 20 inches in a separate area for sick or new animals.

Heat is one of the most important environmental factors which affects rabbit production. Ambient temperature of 30^0 to 32^0 C and above decreases the animals feed intake and increases its water intake. The growth rate is also reduced in growers. Besides, reproductive performance is impaired in adult females with reduced fertility, and temporary sterility in the males. So it is important not to bred rabbits , especially during the summer months, and try to keep the rabbit house as cool as possible during this period to ovoid heat stress in rabbits. To keep the rabbitry cool, use of thatch roofing and hanging of mats or old gunny bags or grass that is at the sided is advisable.

The roof and the sides should be sprinkled with water. One very important point to remember in this regard is that the gunny bags or grass that is should be placed at least one foot away from the sides to facilitate ventilation. This prevents build up of humidity inside the rabbit house.

Nest Box : At the time of kindling, a doe requires a nest box to produce and hide her bunnies. If some type of enclosure is not provided, she will deliver her young on the cage and very often several bunnies will die of exposure and lack of bedding. Remember the bunnies are born hairless and blind and hence require warmth and security of their mother. Therefore, a nest box should be provided for the mother to kindle and keep its litter warm. Usually the nest box is square in shape made of either wood or thick plywood. Its dimensions are 18 x 12 x 9 inches. Its height is 9 inches at the back end and 6 inches in front when viewed from side with the entrance end being only 2 inches high when viewed from the front. At the back end it has roof covering of only 6 inches from where, the sides slope downwards towards the front end so that the mother can enter and exit the nest box during nursing of its young easily. To help keep the nest box dry , drain holes spaced ½ inch apart should be made in the bottom of the nest box.

A kind of circular nest enclosure can be made out of clay. It consists of wide circular clay trough- 12 inches in diameter, 9 inches high and ½ inch thick. A quarter of its circumference is cut in a U shape, 2 inches above the bottom, to enable the doe to enter and exit the enclosure. This places standing upright and not inverted. This cage be indigenously made and is less expensive.

CHAPTER 6

Advances in System of Housing

Definitions

Pen Enclosure for housing rabbits that allows for freedom of movement by rabbits and allows for the provision of a variety of environmental enrichment strategies. Cage Fully enclosed container for housing rabbits which, because of its size, restricts freedom of movement by rabbits and limits the provision of environmental enrichment strategies. Cages are usually constructed from metal or plastic, with solid or mesh sides.

Group Housing in Pens

Advantages of Group Housing in Pens

Principles: The advantages of housing rabbits in groups in pens include:

- Housing in pens provides increased space which allows rabbits freedom of movement to carry out normal activities such as hopping, stretching out, sitting up with ears erect, rearing, and leaping. Physical and psychological well-being is assisted by the opportunity to exercise and explore a complex environment.
- Rabbits are social animals which benefit from the company of others. Housing rabbits in groups in pens allows for social interaction and behaviours such as grooming pen mates, lying together and playing.36, 37 The behavioural repertoire of group housed rabbits is more varied compared to that of singly housed rabbits.
- Pens provide a greater opportunity than cages for the environment to be enriched and made more behaviourally stimulating (for example by the addition of ledges for climbing on and areas for retreating into).
- It is reported that the costs of setting up pens and maintaining rabbits in pens is less than for buying cages and maintaining rabbits in cages. Cost savings may be made in the areas of bedding, cleaning agents, maintenance, energy and labour.
- Health advantages include that sore hocks (pododermatitis) and gastrointestinal hair balls (trichobezoars) are rare or non-existent in penned animals. The incidence of "snuffles" may also be reduced, possibly because of better ventilation than solid walled cages, although an increased incidence of sneezing has been observed.
- Thinning of the bones of the femur and spine because of inactivity may be seen in caged rabbits. This can result in a loss of ability to move normally, fractures, spinal distortions and discomfort from trapping nerves in the spine. There is no evidence of such thinning of the bones, with its attendant complications, occurring in rabbits in pens.

Disadvantages of Group Housing in Pens

Principles: The disadvantages of housing rabbits in groups in pens include:

(i) A high level of stockpersonship in caring for and monitoring rabbits is required. Thus animal carers need to be proficient in recognising all the different aspects of rabbit behaviour. This is so that both health and social interaction can effectively be monitored.

ARRP Guideline 18: Guidelines for the Housing of Rabbits in Scientific Institutions

(ii) Fighting and bullying may occur, especially in mature males. The grouping of rabbits may become unstable – there may be difficulties in reintroducing a rabbit that has been removed for even a short time. In addition, stable groups may become unstable, resulting in aggressive behaviour, for no obvious reason. Fighting and aggressive behaviours may result in severe injuries and stress responses in subordinate animals.

(iii) Rabbits may be difficult to catch in pens.

(iv) Identification of individual rabbits may be more difficult.

(v) Bedding disposal may cause a problem because of its volume.

(vi) Space may be wasted, especially vertical height.

(vii) There is the potential for disease spread. However this has been reported not to be a problem in practice.

(viii) It has been argued that rabbits in pens may be subject to more variables which may affect the interpretation of experimental results. This, however, needs to be weighed against the effects of obtaining data from rabbits, singly housed in cages, which may be physiologically and psychologically abnormal.

Recommendations

Rabbits should be Housed in Groups in Pens

Rabbits that cannot be housed in groups (e.g. intact males or for experimental reasons) should be housed in pens with olfactory, visual and, if possible, physical contact with adjacent rabbits. Rabbits should only be housed in cages with the express permission of the Animal Ethics Committee of the institution on the basis of compelling evidence for the need to use such housing. In such cases, cages should be enriched by methods as described in this document (such as pair housing in double cages). Lack of space or facilities for pens should not be considered sufficient justification for the use of such cages.

Rabbits should not be housed singly in conventional (un-enriched) cages except in exceptional circumstances and with the express permission of the Animal Ethics Committee of the institution on the basis of compelling evidence for the need to use such housing. Such circumstances may, for example, relate to research into the effects of such caging on rabbits. Lack of space or facilities for pens should not be considered sufficient justification for the use of conventional cages.

ARRP Guideline 18: Guidelines for the Housing of Rabbits in Scientific Institutions.

Pen Design and Environment

Pen Construction

Costs of Pens

Principles

(i) The cost of group housing in floor pens is likely to be considerably less than single housing in cages. Savings may be made in areas including cage washers (running, maintenance and servicing), chemicals (descalers and disinfectants) and staff time for cleaning (although some of the staff time saved will need to be spent monitoring and handling rabbits in pens).

Materials for Pens

Principles

(i) A variety of materials may be used, depending on the design of the pen. For example, the pen may occupy a whole room or be adapted from plastic bins made for other purposes or from pens for larger animals. Materials for pen walls may be solid (for example plastic) or be open weave (for example wire mesh) and opaque or transparent / translucent. Opaque solid walls between pens have the advantage of providing additional areas for shelter/ privacy, but the disadvantages, as with all solid walls, of restricting air flow and of restricting vision of surroundings (including of rabbits in adjacent pens and people approaching).

Recommendations

1. Materials used for pen construction should be safe and non-toxic for rabbits.
2. Materials used for pen construction should be easy to clean, water resistant and strong.12
3. For the front of the pen, material that allows rabbits to see out (for example wire mesh) is preferable to opaque material, as this allows rabbits to see who is approaching, which helps to eliminate fear responses.

Floor Area of Pens

Principles

(i) The age and activity level of rabbits is more important in determining space requirements than the weight or size of rabbits. For example, young rabbits need more space than adults for play behaviour.

(ii) The minimum space provided should allow each rabbit to carry out its normal behaviour, including a wide range of locomotory behaviours, such as hopping, leaping, playing, exploring and stretching out. It has been suggested that the minimum space provided should allow rabbits to complete 3 hops in one direction. Adult New Zealand White rabbits have been measured travelling 1.5 - 2.0 meters in three "normal" hops. They will travel much further in three hops when startled.

ARRP Guideline 18: Guidelines for the Housing of Rabbits in Scientific Institutions

(iii) In addition to meeting minimum space requirements for movement, space should be provided to allow the provision of structural complexity and environmental enrichment in pens.

For example, additional space should be provided to accommodate objects such as boxes and pipes, which provide rabbits with retreat / hiding areas.

Table 1: Summary of space requirements

No. of rabbits	Minimum space (m^2) (*ensuring a CLEAR area of $2m^2$ and a minimum length in one direction of 2m)
1	2.00
2	2.00
3	2.25 - 2.40
4	3.00 - 3.20
5	3.75 - 4.00
6	4.50 - 4.80
7	4.75 - 5.05
8	5.00 - 5.30

Recommendations

A minimum CLEAR area of 2.0 square meters should be provided, minimum length in one direction of 2.0 meters should be provided. An overall minimum floor area of 0.75 – 0.8 square meters per rabbit should be provided for groups of up to 6 rabbits. For groups of more than 6 rabbits, space should be allocated at approximately 0.25 square meters per additional rabbit. For a group of 6 – 8 rabbits a minimum floor area of 4.5 – 5.0 square meters should be provided.

Height of Pens

Principles

(i) Normal behaviour of rabbits includes rearing up on their hind legs and sitting up erect with their ears pricked. Such erect stances allow rabbits to investigate sights and sounds. The minimum height provided should allow rabbits to carry out these behaviours.

(ii) Rabbits commonly sit on top of objects (for example ledges and nest boxes), viewing the environment from a raised area, and the minimum height provided should allow for this behaviour.

(iii) Rabbits are capable of jumping very high, and of using objects within a pen as "launching pads" to jump out of pens.

Recommendations

1. If the pen is covered, a minimum height of 100cm should be provided. The minimum height should enable rabbits to sit on top of raised objects.
2. If the pen is uncovered, a minimum height of 1.25 meters should be provided.
3. The proximity and height of objects near the edges of pens should be taken into account in making a pen secure from escape.

Bedding in Pens

Principles

- Bedding ensures a clean, dry, comfortable lying area.
- Ideally bedding should be dust free, free of microbial or parasitic contamination, non-toxic, ammonia binding,

non-traumatic and moisture absorbent. In addition it is desirable for it to be cheap, readily available and easy to use and dispose of.

- The smell of bedding is important - some woods have resinous aromatic oils which affect rabbits' behaviour patterns by altering the olfactory field. There is evidence that rabbits avoid sawdust and wood shavings.
- Straw has the advantage of doubling as a means of environmental enrichment as rabbits play with, nibble, manipulate and burrow in straw. It has the theoretical disadvantage of being a potential source of microbial disease due to contamination.
- To overcome this, straw can be autoclaved, however, this is expensive and creates practical difficulties. It is reported that, in practice, the introduction of disease has not been a problem. A drawback of straw it that it may not be as moisture absorbent as other bedding materials such as wood shavings.
- Shredded paper has the advantages that it is clean, dust free and odourless. However, it has poor absorbency qualities. Rabbits have been shown to choose shredded computer paper over bare concrete floors, sawdust or wood shavings.

Recommendations

1. Bedding should be provided.
2. Straw is the bedding of choice and should be supplied to a depth of at least 5cm.
3. Shredded paper is the next bedding of choice. It should be supplied to a depth of at least 2cm.37 Because of its poor absorbency qualities, it is best to put shredded paper on top of multiple layers of newspaper.
4. Sawdust and wood shavings may be used. They should be supplied to a depth of at least 2cm. If these materials

are used they should not smell strongly and should be made from non-resinated (softwood) timber. Autoclaving may remove odours that offend rabbits.

5. To increase the moisture absorbent properties of bedding, while supplying material that can be manipulated and played with, a layer of wood shavings or sawdust may be used under straw.

Cleaning of Pens

Principles

(i) Rabbits in pens stake out a toilet area - they urinate and defecate in one corner, although faecal pellets may be scattered with the movement of rabbits in the pen.

(ii) A balance needs to be struck between the human perception of the need for cleanliness and the level of disturbance to rabbits. A sense of smell appears to be very important to rabbits - social interactions related to sex, age, reproductive status, individual identity, hierarchy and mother/young relationships may be communicated via the olfactory environment. Scent signals arise from urine and secretions from specialised scent glands (chin, inguinal, anal and Harderian glands). It has been suggested that urination and scent marking carried out by rabbits in clean cages are attempts to restore an "optimum odour field".

Recommendations

1. Frequency of cleaning will be influenced by factors including the type and depth of bedding, stocking rates and the efficiency of air exchange in decreasing levels of noxious gases.

2. Cleaning the toilet area (for example 3 or 4 times a week) is usually required in between completely changing the pen bedding.

3. Two weeks is the longest recommended interval between complete changes of bedding.
4. If a system of shredded paper over layers of newspaper is used, the newspaper needs to be changed at least twice weekly.
5. A pen should ideally be washed and disinfected (with an odourless disinfectant) at least every month.
6. Pens and rooms should be completely cleaned every time a room is cleared and new stock brought in. The floor should be washed and an odourless disinfectant used.

Rabbit Care and Management in Pens and Cages

Management of Rabbits in Groups

Principles

I. The success of group housing rests largely with the skill and enthusiasm of the animal carers. Carers must be able to monitor and assess rabbit behaviour and implement management strategies accordingly.

II. Some breeds are more aggressive than others. Dutch rabbits are usually more aggressive than New Zealand White rabbits, and are generally unsuitable for group housing. Lop rabbits are usually more docile than New Zealand Whites.37 However, a stable group of rabbits cannot be guaranteed by selecting less aggressive breeds. The occurrence of aggression, which rabbits will be involved in aggressive encounters, and the severity and outcome of such encounters, can be difficult to predict.

III. It is best to establish groups of rabbits when they are young (around the time of weaning), and at least before they reach puberty (which may begin as early as 2.5 months of age). Ideally littermates should be used, although separate litters of similar age can be grouped. Large weight or age differences can result in aggression. All-female groups are more likely to be stable. Sexes should not be mixed (apart from breeding groups).

IV. Males can be housed in groups until they reach 3 - 4 months of age. However, mature male rabbits fight and this fighting can result in severe injuries. The lack of space in pens for rabbits to escape aggressive encounters makes it impossible to keep mature male rabbits in groups in pens.34 After about 3 - 4 months of age, housing intact males together in the expectation that they will be separated before fighting starts is unsafe for the rabbits, as the outbreak of severe aggressive encounters is difficult to predict and may occur without warning.

V. Castration has been shown to be a viable alternative for housing males together. Castration should be carried out at about 10 weeks of age - i.e. before aggressive behaviour is shown.

VI. Some individuals may be highly aggressive. In such cases it may be necessary to remove the dominant or subordinate rabbit temporarily or permanently.

VII. It is important that rabbits are provided with sufficient space and objects to assist them to escape and hide from their aggressors. Objects such as boxes, pipes, ledges and vertical barriers provide means for hiding and escape. Sufficient objects should be provided to eliminate competition for such items.

VIII. Housing more than 6 - 8 rabbits together creates difficulties in developing and maintaining a stable hierarchical system. Limiting groups to this number also assists in monitoring animals for signs of bullying and ill health.

IX. It is very important to keep the composition of a group stable. Where individual rabbits need to temporarily be separated from a group, they should be housed so that visual contact between the individual and the group can be maintained. This helps to ensure that they will be readily recognised and accepted as familiar members of the group when returned.

X. In a group where a rabbit has been removed (for example for a scientific procedure) and then returned, rabbits should be closely monitored for signs of aggression (for example, chasing, fighting, fight wounds).

Recommendations

1. For monitoring the behaviour of rabbits in pens and implementing rabbit management strategies, animal carers should have suitable qualifications and experience. Their skill and enthusiasm are important factors in the successful housing of rabbits in groups in pens.

2. Wherever possible, rabbits in groups in pens should be of similar age and weight, be the same sex (preferably female), and be litter mates or mixed together at an early age (around weaning).

3. Mature (un-castrated) male rabbits should not be housed together.

4. Sufficient space and objects (such as boxes, pipes, ledges and vertical barriers) should be provided to assist rabbits to escape and hide from aggressors. Sufficient objects should also be provided to eliminate competition for these items.

5. The number of rabbits in a group should be limited to 6-8 (with the exception of breeding groups).

6. Where individual rabbits need temporarily to be separated from a group, they should be housed so that visual contact between the individual and the group can be maintained, to assist with reintroduction into the group.

2 Regrouping / Establishing Groups from Caged Rabbits

Principles

(i) It is possible to regroup rabbits, but this requires particular care and intensive monitoring. It is important that rabbits are placed in a fresh neutral area to avoid home territory for any animals.

Additional measures include

- Providing hiding places and breaking up clear areas so that rabbits can escape from each other;
- Placing wire partitions between pens to allow familiarisation between rabbits before they are mixed;
- Scattering faecal pellets and urine soaked litter from each rabbit in the new pen;
- Scattering food in the new pen to encourage foraging;
- Having the usual carer handle rabbits together in small groups before mixing.

(ii) A method for regrouping rabbits using sedation has been described by Love 91. In this method rabbits are sedated (fentanyl and droperidol) and placed in a small holding cage, making sure pen mates are not next to each other. The important part of this process is that rabbits chin and rub against each other as they are recovering from sedation, thus spreading recognizable smells. Rabbits are then placed in a pen not previously used by any of the rabbits. Rabbits may display behaviours such as chasing but usually settle within 24 hours.

(iii) Moving rabbits from cage housing to pens also presents particular problems. Rabbits that have been housed in cages for 6 months or more are prone to fighting (due to lack of social experience) and bone fractures due to osteoporosis. Methods to assist in group housing such animals in pens include providing increasing daily periods of exercise individually and then in small groups, then housing in the group pen during the day and individually in cages at night and finally housing in the group pen. Even with such methods, signs of aggression need to be looked for. Groups may become unstable and may require the removal of individuals.

Recommendations

1. Particular care should be exercised in regrouping rabbits or in moving rabbits from cage housing to group housing in pens.
2. Catching / Handling of Rabbits in Pens and Cages

Principles

(i) Because pens provide room to escape, catching rabbits in pens may be difficult. Anticipation of what is to happen once caught plays a major role in a rabbit's behaviour. Therefore steps should be taken to reduce the stress of procedures and to catch and handle rabbits without conducting other procedures.34 Animal carers should also spend time with the rabbits, which will facilitate catching.

(ii) Rabbits show a reduction in fearfulness to handlers after repeated approach and handling. The use of handling and approach programmes may help reduce general emotional reactivity (not just fear of humans) and strengthen the human-animal bond.

(iii) A quiet approach should be taken. Rabbits will usually retreat to a darkened hiding place from where they may be picked up.

(iv) In one study, rabbits were trained to accept a procedure (oral administration of an antibiotic solution) without restraint when given positive reinforcement (drug delivered in a sugar coated syringe after a period of training by administering a sugar solution).

Recommendations

1. To assist in catching rabbits, a quiet approach should be taken.
2. To assist in catching rabbits, a darkened retreat area from which rabbits may be picked up should be provided. For example, provide plastic pipes of about 20 - 25cm diameter, which can be upended to catch rabbits when they enter them.
3. It is important that rabbits are caught and handled on a regular basis (ideally daily and at least weekly) to facilitate ease of catching and to reduce fearfulness. Such catching and handling should be carried out in addition to catching and handling rabbits in order to conduct procedures.
4. The training and rewarding of rabbits using positive reinforcement or "treats" should be considered when performing procedures on rabbits. This is likely to reduce the stress on rabbits and increase their co-operation.

Identification of Rabbits in Pens and Cages

Principles

(i) Clause 5.6.1 of the Australian Code of Practice for the Care and Use of Animals for Scientific Purposes states: The method of identification of individual animals must be that which causes the least distress within the context of the research proposal and the least interference with the normal functioning of the animal.

(ii) Ideally methods of identification should not be painful, not cause adverse reactions, not be uncomfortable and not likely to catch or tear out.

Recommendations

1. The least invasive method of identification should be used that is compatible with the use of rabbits in the institution.

2. Coat colour and pattern is useful for the identification of individuals in some breeds of rabbits.
3. Dyes including permanent wool dyes, agricultural sheep sprays, fuchsin, acriflavine or gentian violet may be used. These need to be reapplied at intervals, as the dye fades or the rabbit moults.
4. Xylene free permanent markers may be used on ears and fur. Generally these need to be refreshed every 3 weeks.37 The use of these markers on the inside of the ears is effective and different colours can be used for coding.
5. Fur clipping may be used but needs to be carried out frequently.
6. Microchips and ear tattoos may be used for permanent identification. Note there is some transitory pain associated with applying these forms of identification. Microchips can be inserted at 6 weeks of age. Anaesthesia or sedation and analgesia should be used in applying tattoos.
7. Leg rings, fitted over the hock at about 6 weeks of age may be used. However, they need to be checked frequently while rabbits are growing (at least weekly) to ensure they do not become too tight.
8. Studs or ear tags should not be used as they may tear out, they may cause infections leading to the growth of fibrous tissue as a chronic inflammatory response, and they are painful to insert.
9. Collars should not be used as they may be chewed off by other rabbits, they may drop off and rabbits may catch their legs in them.
10. Ear notching must not be used.

Food and Water for Rabbits in Pens and Cages

Principles

(i) Dried pelleted diets have the advantages of being standardised with little variation. However, these diets are monotonous and their repetitive presentation without supplements (such as hay and green feed) may compromise the well-being of rabbits.

(ii) Scattering food reduces boredom by encouraging rabbits to forage. This reduces the amount of time available for fighting and helps to prevent obesity.

(iii) High fibre feed such as hay helps prevent diarrhoea and trichobezoars (hairballs). A diet with around 18 - 25% crude fibre is suggested. Hay also provides something to manipulate and play with.

Recommendations

1. A nutritionally adequate diet should be provided for rabbits.
2. It is recommended that more than one source of food and water be provided to reduce the possibility of aggressive competition.

   2 Dried pelleted diets should be fed at around the amount of 60 - 80 g/kg/day depending on factors such as the age of rabbits and their opportunities for exercise. To relieve boredom, more than one formulation may be fed.
3. Hay should be fed on a daily basis.
4. For variety, pelleted food should be supplemented with items such as fruit and vegetables, corn, barley, oats and soya beans. Fruit and vegetables should be washed to reduce the risk of introduction of disease.
5. Food may be spread out to encourage rabbits to forage. Food may be scattered or placed in small piles around the pen. Fouled or uneaten food should be removed daily.
6. If a restricted diet is required to be fed, it should be provided in the late afternoon (which has been shown to reduce the frequency of abnormal behaviours in caged rabbits).

7. A plentiful supply of clean water should always be available. Various systems to supply water may be used. If automatic watering systems or water bottles are used, care should be taken to avoid leakage and overflows. Open water systems (such as chicken water hoppers) should be raised up to prevent water being contaminated with bedding and faeces.

6 Monitoring of Rabbits in Pens and Cages

Principles

(i) A high standard of animal care is crucial for the success of housing rabbits in groups in pens. Rabbits must be monitored both for health and for social interactions within the group.

(ii) In cages normal behaviours are difficult to assess (because it is difficult or impossible for rabbits to carry out these behaviours) and changes in food and water intake are often the only early indicator of illness.

Recommendations

1. Rabbits should be monitored by observation at least daily.
2. Weekly health checks of all rabbits should be carried out by animal carers (in addition to daily observation)
3. Animal carers should be aware of the normal behaviour of rabbits and of the individuals within a group and observe for deviations from normal.
4. In particular, subordinate rabbits should be monitored for signs of bullying (which may, for example, result in fight wounds or denial of access to food or water).
5. Rabbits that give cause for concern (either excessively aggressive or timid) may need to be removed from a group.
6. To ensure that individuals can be adequately monitored, group sizes should not exceed 6- 8 rabbits.

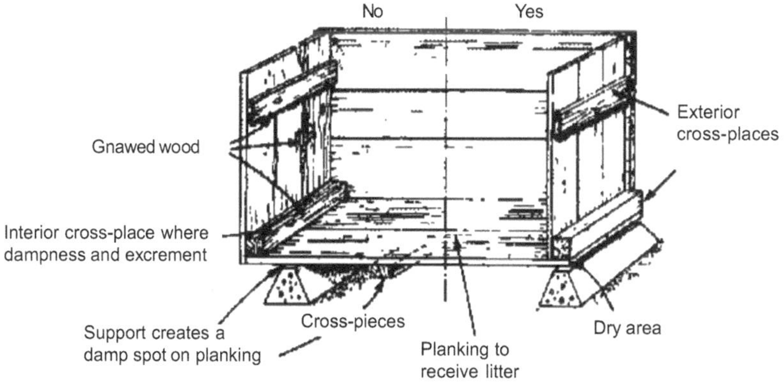

Fig. 1 : Examples of correct and incorrect cage assembly, fostering good hygiene and resistance (a)

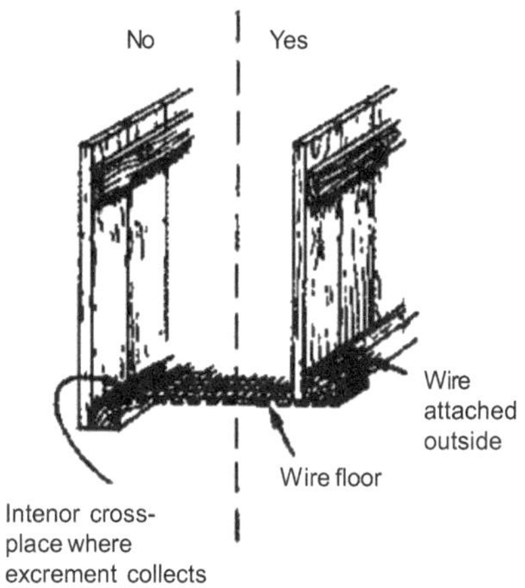

Fig. 2 : Examples of correct and incorrect cage assembly, fostering good hygiene and resistance (b)

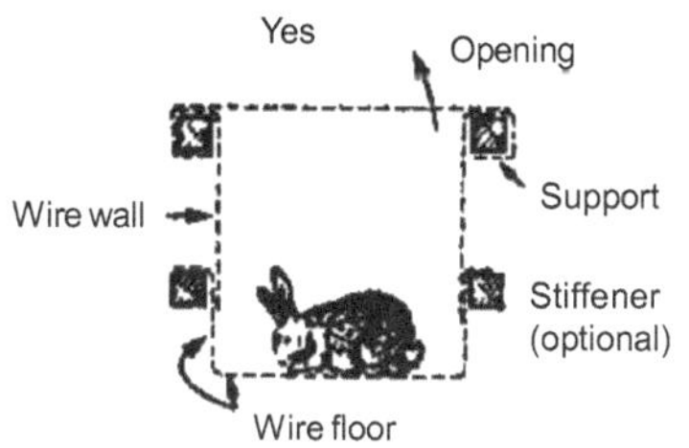

Fig. 3 : Examples of correct and incorrect cage assembly, fostering good hygiene and resistance (c)

Note: a = wooden cage with litter, b = wooden cage with wire floor; c = suspended wire cage.

Watering

A permanent dispenser of clean water is an essential item in each cage, wherever rabbits are not fed green forage alone.

Using old cans or glass or earthenware pots as drinkers can create a hygiene problem. Rabbits tend to soil their water, especially if they are reared on straw litter. The drinkers should be fastened so that the rabbits cannot overturn them and so that the breeder can easily clean and refill them once or twice a day.

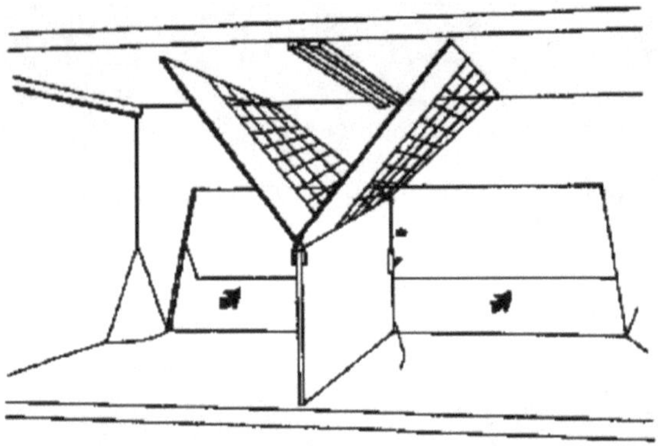

Figure 4 Swinging rear wall in concrete hutch for waste removal (a)

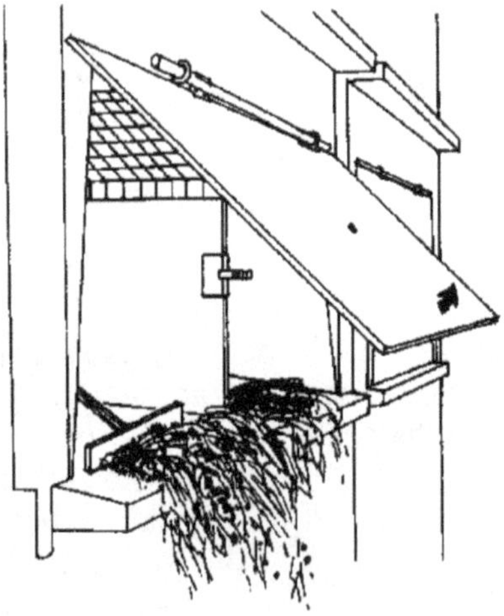

Figure 4 Swinging rear wall in concrete hutch for waste removal (b)
Note: a = front view (note forage rack, here shared by two cages); b = rear view.
Source: Thébault, Rougeot and Bonnet, 1981

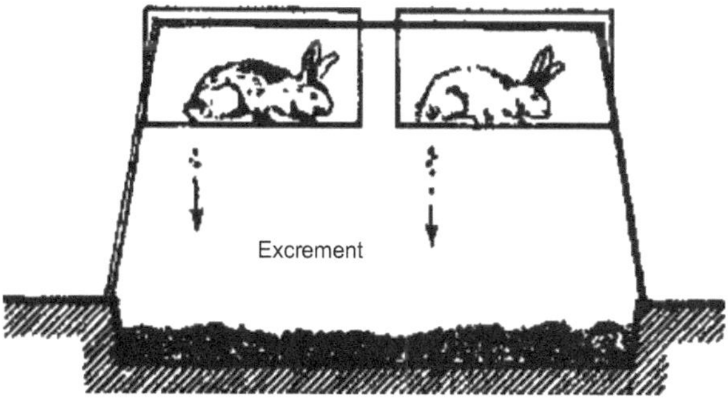

Figure 5 Four systems for using wire cages (a)

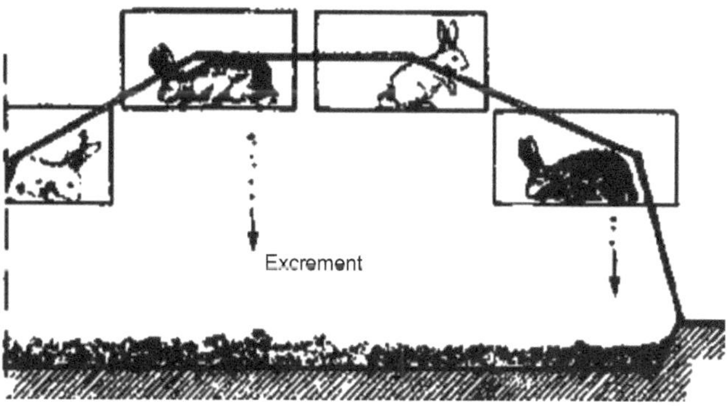

Figure 5 Four systems for using wire cages (b)

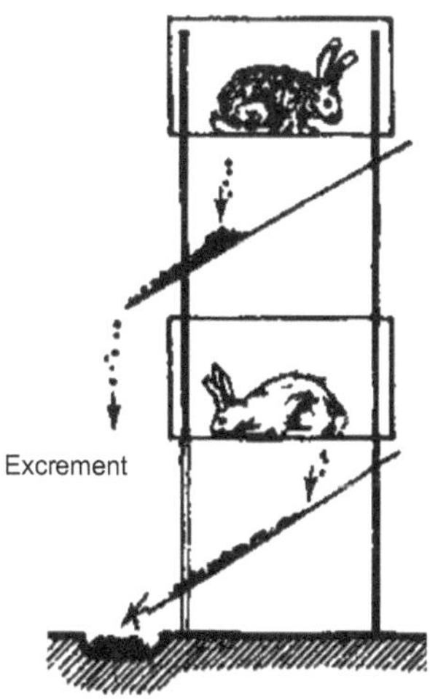

Figure 5 Four systems for using wire cages (c)

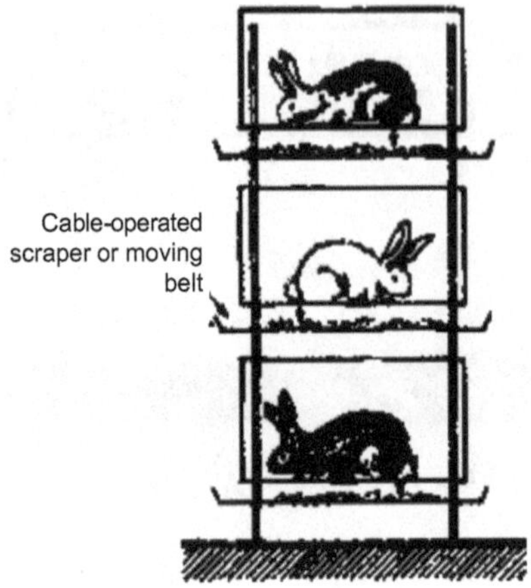

Figure 5 Four systems for using wire cages (d)
Note: a = flat-deck; b = Californian; c = inclined-slope battery; d = compact battery.

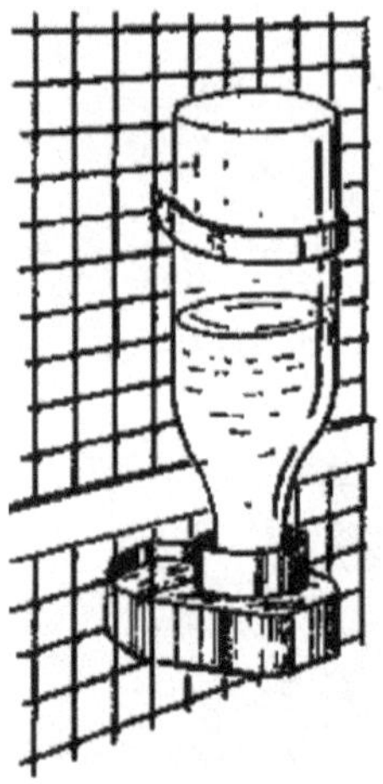

Figure 6 Inverted water-bottle drinker

The Nest Box

The nest box should be considered one of the most important items of equipment in rabbit production. It directly affects the viability of the young in the pre-weaning stage, which is the high-risk mortality period (15 to 40 percent of live born rabbits). The job of the nest box is to reproduce conditions in the burrow of a wild doe and protect the young against attacks from the outside environment so that they can get through the first few difficult days of life in optimal comfort. To do this the nest box must:

- Allow the doe to kindle and nurse her young in comfort;
- Keep the young in a healthy, clean environment;
- Prevent dampness from the animals' urine;
- Keep the young together in cold weather and help them maintain a constant temperature close to 30° to 35°C in the middle of the nest;
- In hot weather, allow the doe to scatter the nest so that the young can adapt to the temperature;
- Keep the young from leaving the nest too early and make it easy for them to get back if they do get out;
- Allow thc producer to monitor the litter, remove any dead animals, introduce baby rabbits to be fostered and change bedding material easily, without disturbing the doe and the young.

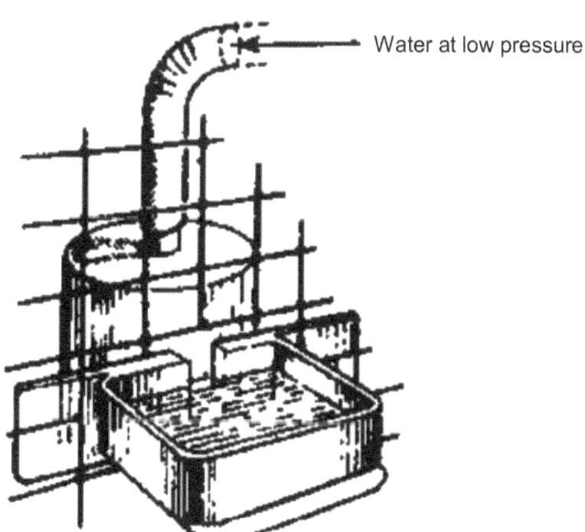

Figure 7 Automatic drinkers (Open drinker)

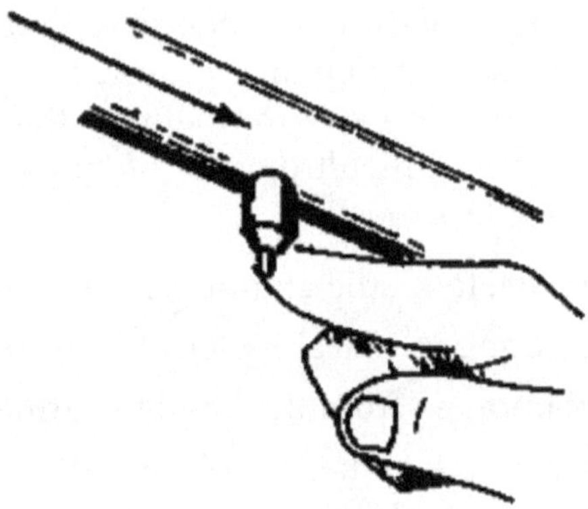

Figure 8 Automatic drinkers (Open drinker)

Figure 9 Drinker made from a nipple in plastic bottle (a)

Figure 10 Clay drinker used as inverted water bottle (b)

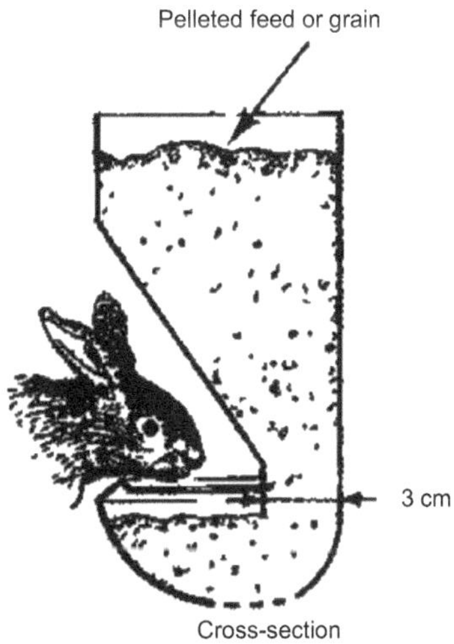

Figure 11 Feed hopper (a)

Source: Finzi and Amici, 1992. FIGURE 11 Feed hopper (b)

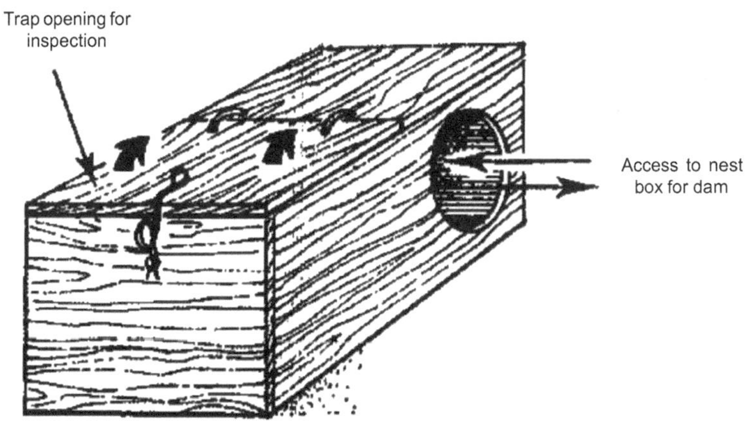

Figure 12 Design for a nest box

Note : In hot climates the top of the box can be made of wire mesh.

Buildings: In Constant Hot Climates

In countries where the climate is hot but fairly constant (mean minima and maxima between 20° and 30°C) closed buildings are not really necessary. All that is needed is to protect the rabbits against the weather. If the cages are of wood or concrete (solid walls) it may be enough to roof each hutch, as shown in Figure 34. A roof should keep off rain and also heat from direct sunlight. The hutches can also be placed under trees big enough to shade them all day long. A roof should overhang enough to keep water out on rainy, windy days. The hutches should face away from the prevailing winds.

Wire-mesh cages can be grouped under a common insulating roof. This system, illustrated in Figure 35, was first tried in California. It is satisfactory provided the roofs overhang far enough at the sides to protect the animals properly. A hedge or fence around the roof structure is useful in protecting the rabbits from strong winds, and from predators.

In Variable Hot Climates

In such climates the rabbits must be reared either on litter in hutches out in the open, or in cages placed inside a building which will serve as a buffer against the heat. Very satisfactory results have been obtained in Burkina Faso with buildings constructed with local Palmyra (Borassus aethiopium), and a straw roof. The temperature in a building like this is more constant than in a more costly one made with concrete perpend.

At the Irapuato National Rabbit Centre in Mexico, solid buildings are generally left open in the front during the day and at night the shutters are closed to offset the drop in outside temperature. A daily temperature range of 20°C is common in the region. These shutters also make it possible to ventilate the interior during the daytime; they can be opened to suit the wind direction and regulated to respect the air-flow standards mentioned at the start of this chapter.

In some dry tropical regions of Africa where wood is scarce, producers have made satisfactory housing by building small round huts of unbaked earth bricks covered with straw, used for both cage and housing. Litter changing is often quite a problem with this sort of construction, however. The floor should slope slightly and be off the ground. Parasitism can be partially controlled by demolishing the hut every year and rebuilding it a few meters away. Such housing is thus only suitable for backyard rabbitries in which labor is not a problem

Unconventional Housing

The usual techniques for cages and buildings known to give reliable results in all climates have been described so far. This does not rule out other practical solutions, some examples of which are given below.

Underground Rabbitries

In the southern parts of Tunisia and Algeria, breeders traditionally rear rabbits in a dry "well" 1.5 to 2 m deep (Finzi, Tani and Scappini, 1988). Breeders first dig the well and then lower the rabbits that will breed a colony, building burrows at the bottom of the well. These are used by does as nests, reflecting wild rabbit burrows. The breeder simply throws down fodder, which can occasion significant waste. In more elaborate rabbitries, the breeder digs a sloping tunnel from the bottom up to ground level where it emerges into a small pen. The feed is set in the pen and the rabbits come for it at will (usually at night). A trapdoor in a corner of the pen allows the rabbits to be caught. Of course this system can only work in countries where it hardly ever rains and the ground remains dry down to 1.5 to 2 m. Another drawback is uncontrolled breeding and the breeder may easily maintain totally unproductive rabbits for long periods. Predator control is virtually impossible as well, particularly for rats.

Finzi (1992) describes another underground pen for group rearing, the result of field observations and experiments, shown in Figure 36. Note the simple predator control and rabbit shelter concepts.

Cage Rearing

A system of cages using broad cement channels (0.8 to 1 m wide) laid horizontally has been described in Spain by Contera (1991). The rabbits live on a wire-mesh floor slightly narrower than the channels, and the droppings fall into the channel. The cage/shelters are conventionally equipped. In the hottest hours of summer, systematically sprinkling the outer walls of the channels cools the temperature compared with the outside, the water evaporating through the fairly porous cement of the channels.

Another heatproof device was described by De Lazzer and Finzi (1992): a system of dual-zone cages. Outside is the conventional wire-mesh cage with the feed racks and inside an "area" of equal volume buried beneath a layer of thick earth but accessible to the breeder through a trap, with the two connected by a 20 cm fibrocement tube. These authors have reconstituted a living area for the fattening or nursery rabbit(s) in the cage that resembles the living space of wild rabbits. During the hot hours of the day, or when there is a disturbance (or to kindle), the rabbits stay underground. When hungry or thirsty, they move into the wire-mesh cage. Experience has shown that the animals always use the outside as a latrine area. The technical results obtained by these authors in one year indicate a productivity wholly comparable with conventional closed rearing in cages, but at lower cost. No information is available on the labor required to build the system.

Uses for Waste

In every type of rabbit production unit the producer has to remove the excrement and waste from the rabbitry (straw litter and droppings which pile up under cages). These can be put to good use on the farm. The amounts and composition of waste vary according to housing and feeding conditions.

Rabbits eating balanced concentrate feeds and raised on mesh floors produce about 25 to 400 g of faeces and 0.5 to 0.8 liter of urine per mother-cage a day, depending on production intensively. This waste is much richer in nutrients than ordinary farm manure. In fact, farm manure contains only 0.4 to 0.6 percent of each of the main fertilizer components: N, P2O5 and K2O.

The composition of the waste varies with the type of rabbit. A comparison of the figures in Tables shows a greater risk for nitrogen and phosphorus losses during storage than for other elements.

The average composition of the manure of rabbits reared on litter depends partly on the kind of feed but mostly on the kind and amount of litter used. If well preserved, the waste collected weekly will contain the nutrients in the faeces, part of those in the urine and those in the litter. Fertilizer "production" is therefore at least equal to that in a rabbitry not using litter.

Figure 13 Wire-mesh cages under a common roof

Figure 14 Two-zone cage: wire-mesh (at left) and underground (at right). The breeder has access through the corrugated top laid over the underground area

Source : De Lazzer and Finzi, 1992.

CHAPTER 7

Nutrition and Feeding of Rabbits

Rabbits are produced for meat, wool and research and as pets or for a hobby. Meat production is the most important commercial aspect of rabbitry. The domestic rabbit is primarily herbivorous and will consume most types of grains, oil seed cakes, brans, greens and hay. Diets provided, whether home grown or commercially prepared, consist almost entirely of ingredients from plant sources. Although, the producers / farmers mostly rely on home grown feeds, a good portion of the rabbit feed presently used is commercial pelleted feed. Since the rabbit can utilize a certain amount of forage, it has a place in food production by making use of some non-competitive feeds.

Rabbit's digestive system resembles to that of a pig and horse. Habitually rabbits practice coprophagy, sometimes referred to as pseudo rumination. It refers to the production of two kinds of faecal matter, one hard and one soft, the latter being consumed directly from the anus as it is excreted at night. This practice begins in rabbits mostly after they begin eating solid feed at about 4 to 6 weeks of age but is not practiced by germ-free rabbits. Fermentation in the large intestine and practice of coprophagy probably provides necessary amounts of most B complex vitamins, some quantity of microbial (bacterial) protein and may permit further digestion of some nutrients by multiple passage through the digestive tract. The high digestibility of forage protein in rabbits may be partially due to coprophagy. Rabbits have a number of characteristics that suggest that they can play a major role to complement the growing of crops to make most efficient use of resources, particularly in developing countries. Some of these special attributes of rabbit include rapid growth rate, high reproductive potential, use of non-competitive feeds and small body size.

Feeding

The rabbit is a herbivorous animal, and eats feeds of plant origin like green or dry forages, vegetables, roots, tubers, seeds, tender stems and leaves. It can eat both fresh and dry portions of a variety of plants and grasses and as well as all kinds of kitchen scraps and waste. Although they can live and reproduce on these type of feed they nevertheless need some concentrate feed containing grain, bran, groundnut cake etc., for efficient production.

Rabbits are generally divided into 4 classes based on their dietary needs:

a) Growing and fattening rabbits.

b) Resting bucks, non-pregnant and non-lactating does

c) Pregnant does.

d) Lactating does with litters.

Each of these classes should be fed a different ration in varied amounts.

Planned feeding can be done if home produced feeding stuffs are supplemented with concentrate feed to ensure a properly balanced ration. These days in developed countries rabbits are fed commercially prepared pelleted feed made up of grains, oil seeds, Lucerene hay, bran and certain supplements than just simple mash. Pelleted feed is a sped form of feed made from properly ground and mixed feed ingredients which at first is in a mash (coarse powder) and then this mash is processed with the help of steam or molasses and forced through 4 mm diameter holes making the feed into solid cylindrical pieces of about ¼ or 1/8 inch long, which is a convenient size for the rabbit to piece easily and eat. Rations fed to rabbits kept on wire mesh floors are always pelleted to prevent selection of a feed ingredient and consequent wastage of food.

They prefer a pelleted diet to one in a mash form the major disadvantage of pelleting is the added cost. Non- pelleted or mash type diets have not given encouraging results as pellets have. Adding 5 to 10 per cent molasses to mash diet is more advantageous than just feeding plain mash.

For commercial feed, both pellets and mash, the farmer would have to spend a large amount of money on feed for his rabbits, which is seldom possible. The farmer must therefore, find feed for his rabbits from local sources. For this he can have recourse to large number of plants, either wild (grasses, pluses, shrubs etc.) gathered during weeding; other cultivated plants like maize, jowar, etc.; different kinds of grasses like napier, para, guinea; tree leaves, leafy vegetables; various agricultural by-products like sugarcane tops, various straws, hays and mill by-products like wheat bran, rice bran and even broken rice etc.

Rabbits raised on backyard can be fed a variety of home grown grains, legume hays, grasses and kitchen wastes. In general, rabbits can eat almost any food that is not sour, spoiled or greasy. Rabbits can be fed potatoes, potato peelings, large amounts of cabbage and cauliflower leaves, vegetable trimmings, fruit peelings, bread crusts, sun dried left over rice, etc. However, it must be mentioned that this type of feeding is not satisfactory because the rabbits grow slowly and take longer time to reach the slaughter weight of 2.5 kg. Home grown feeds when supplemented with concentrate feed is satisfactory and gives better results.

Breeding does and growers need more protein, minerals and energy than rabbits that are just being maintained. The concentrate feed should contain crude protein for the various categories of rabbits as follows:

a) Maintenance = 12% C.P.
b) Pregnant does = 15% C.P.
c) Normal growth and fattening = 16% C.P.
d) Lactating does = 17% C.P.

Resting bucks and does (rabbits which are not being used for breeding) need about 120 grams of concentrate feed per day and pregnant does about 240 grams per day. Growing and fattening rabbits (after weaning to slaughter or upto 3½ to

4 months of age) about 60 to 120 grams per day. The feed should gradually be increased, as the rabbits grow from the time of weaning at 6 weeks of age, at the rate of 10 grams per week till they attain 12 weeks of age. There after, they should be fed at the rate of 120 grams per day. Lactating does with a litter of about 6 young ones need 350 to 450 grams each day till the litter is weaned and there after the feed should be reduced to 120 grams per day.

A farmer may mix his own concentrate feed if suitable ingredients are available. Some common formulae are:

Ingredient	Formulae(%)		
	I	II	III
Ground Lucerne hay	20	30	40
Ground maize	35	37	40
Wheat bran	30	20	10
Ground groundnut cake	12	10	7
Mineral mixture	2	2	2
Ground salt	1	1	1
	100	**100**	**100**

An all purpose concentrates feed can be made by the following ingredients:

Ingredient	Quantity(g)
Ground lucerene or Berseem hay	200
Ground maize	250
Wheat or Rice bran	280
Ground groundnut cake	140
Bone meal	60
Fish meal	40
Shell grit	20
Salt	10
	1000 g or 1 kg

Forages are essential for their roughage and vitamin value. In addition, forages greatly economise the amount of concentrate feed. Some kinds of forages that can be fed rabbits are:

Legumes : Groundnut, cowpea, Soyabean, Siratro, Field bean, Potato leaves, Desmoduim, Stylosanthes, Velvet bean,

Gaur, Berseem, Lucerne, Horse gram, Bengal gram, Hedge lucerne etc.

Grasses : Guinea, Para, Napier, Corn leaves, Bamboo leaves, Rhodes, Dinanath, Kancha, Sehima, Spear, Anjan, Dhub, Jowar, Bajra, Oats, Sugarcane tops etc.

Rabbits are produced for meat, wool and research and as pets or for a hobby. Meat production is the most important commercial aspect of rabbitry. The domestic rabbit is primarily herbivorous and will consume most types of grains, oil seed cakes, brans, greens and hay. Diets provided, whether home grown or commercially prepared, consist almost entirely of ingredients from plant sources. Although, the producers / farmers mostly rely on home grown feeds, a good portion of the rabbit feed presently used is commercial pelleted feed. Since the rabbit can utilize a certain amount of forage, it has a place in food production by making use of some non-competitive feeds.

Rabbit's digestive system resembles to that of a pig and horse. Habitually rabbits practice coprophagy, sometimes referred to as pseudo-rumination. It refers to the production of two kinds of faecal matter, one hard and one soft, the latter being consumed directly from the anus as it is excreted at night. This practice begins in rabbits mostly after they begin eating solid feed at about 4 to 6 weeks of age but is not practiced by germ-free rabbits. Fermentation in the large intestine and practice of coprophagy probably provides necessary amounts of most B complex vitamins, some quantity of microbial (bacterial) protein and may permit further digestion of some nutrients by multiple passage through the digestive tract. The high digestibility of forage protein in rabbits may be partially due to coprophagy. Rabbits have a number of characteristics that suggest that they can play a major role to complement the growing of crops to make most efficient use of resources, particularly in developing countries. Some of these special attributes of rabbit include rapid growth rate, high reproductive potential, use of non-competitive feeds and small body size.

Nutrient Requirements

Nutrients can be defined as dietary essentials for one or more species of animals. Nutrition of rabbit is provided through supplementation of dietary essentials in the form of proteins, carbohydrates, lipids, minerals and vitamins. Nutrients are required for sustenance and productivity of rabbits.

Proteins

Proteins are essential components of animal tissue. Muscle tissue is largely protein. All enzymes, which facilitate chemical reactions in the body are proteins. Some hormones are proteins or protein derivatives. Blood proteins have a number of essential functions, including nutrient transport and regulation of fluid balance. The immune system is based on proteinaceous antibodies. Cell membranes contain protein. Thus proteins are basic to animal life. Most of the productive functions of livestock involve protein formation or secretion, including meat, milk, wool, fur etc.

The importance of protein quality in rabbit nutrition is well recognized. For rapid growth, rabbits are dependent upon adequate quantities of dietary essential amino acids. Bacterial protein synthesis in caecum has been demonstrated, but this protein, obtained by means of coprophagy, apparently does not make a large contribution to the essential amino acid needs of the young rabbit. Dependence on dietary essential amino acids implies that non-protein nitrogen sources would not be useful to rabbits.

In contrast to other simple stomached animals, such as swine and poultry the rabbit is able to utilize efficiently the protein in forage plants. For example, in pig the digestibility of protein in alfalfa meal is less than 50%, while in rabbit it is about 75%. Thus, it is feasible to utilize considerable amounts of alfalfa in rabbit diets. The value of various protein supplements for rabbits has been studied. For example fishmeal and soybean meal protein reported to support greater growth than peanut or groundnut meal and sesame meal. The variation in growth response to different protein supplements is largely as a consequence of their amino acid composition.

Carbohydrates

Carbohydrates are of two types viz., non- fibrous and fibrous. The non-fibrous types, such as starch and sugars, are readily utilized by animals as energy sources. The fibrous fraction of feeds corresponds to the structural carbohydrates of plant material. In the past, this was measured as crude fiber. Recently the terms cell wall constituents (CWC) and acid detergent fiber (ADF) have become widely used. The fraction CWC consists of hemicellulose, cellulose, lignin and silica, while ADF consists of cellulose, lignin and silica. A high degree of variability was noticed in the digestibility coefficients for fiber fractions. Feeds high in cellulose & lignin have fiber digestibility of less than 15% but on the contrary for non-lignified materials digestibility could be as high as 60%.

In spite of the fact that crude fiber does not serve as an efficient energy source for rabbits, there is evidence that dietary fibre exerts beneficial effects since the fibre levels below 12 % may promote diarrhoea. Reduction in fiber content in rabbit diet may encourage pulling of fur / hairs from each other. This causes depreciation of the value of fur; on the other hand ingestion of excess hairs may result in the production of hairball in the digestive tract. Herd bucks, pregnant does and does with litter may require 14-27% fiber while young rabbits in growing stages and lactating does should be fed with diets having 10-12% fiber content. Fiber content of the diet also has a strong influence on maternal behavior especially nest quality and associated anomalies and there upon the rate of survival of young ones.

Lipids

More commonly known as fats and oils, lipids function mainly as a source of energy. Common sources for inclusion in the ration of rabbits are vegetable oils, since they are considered superior in terms of more digestibility. It is suggested to incorporate fat in diet of the rabbit at the level of 2-5% as a common practice. The effect of dietary lipids (saturated & unsaturated fat, cholesterol) on human health (Atherosclerosis) is a controversial issue, particularly with respect to meat consumption. Certain aspects of this issue have relevance to the nutritional value of rabbit meat (designated as lean) in the human diet.

Essential fatty acid deficiency in rabbits causes reduction in growth rate, loss of hair and changes in male reproductive system including degenerative changes in the seminiferous tubules, impaired sperm development and decreased accessory gland weights.

Mineral Elements

Minerals are inorganic substances. They can be subdivided into two groups namely macro minerals and trace minerals / elements. The macro minerals, which include calcium, phosphorus, sodium, potassium, chlorine, magnesium and sulphur are required in relatively large quantities. The trace elements, including copper, iron, iodine, manganese, zinc, selenium are needed in very small quantities, often expressed as parts per million (ppm or mg / kg).

Calcium and phosphorous are major constituents of bone. In addition calcium has metabolic roles in blood clotting, in controlling excitability of nerve and muscle tissue and maintenance of acid base equilibrium, while phosphorous is a component of such vital cellular constituents as ATP, DNA, RNA and phospholipids. Trace elements as constituents or activators of enzymes are considered essential (e.g.: selenium, copper, iron and zinc) for normal metabolism.

Vitamins

Vitamins have got definite role to play in terms of metabolism in rabbits. Inadequate supply of it may produce deficiency syndrome in rabbits. The symptoms vary depending on the specific vitamin concerned. Vitamins are classified into two groups: fat soluble and water-soluble. The fat-soluble vitamins are A, D, E and K, while the B- complex and vitamin C are the water-soluble group. The importance of B-complex vitamins in rabbits is significant. The water-soluble vitamins are synthesized by caecal microbes. Rabbits used to replenish B complex vitamins from feaces through their coprophagic activity.

The quantitative need of rabbits for many of the known fat soluble and water soluble vitamins has been determined, but quantitative needs have been estimated for only a very few. It is generally recommended to add vitamin mixture containing at least vitamin A, D and E at moderate levels in rabbit diet.

Water

Water is essential for their normal health and productivity. The requirement of water depends on the climatic condition and physiological status of the animal. About 300 ml to 1500 ml of fresh water is to be provided daily.

Dietary Ingredients

Rabbits consume many kinds of feeds satisfactorily. The feeds selected for use should be determined by relative costs and regional availability.

Classification

1) Green or Succulents Feeds

Growing pasture plants, root crops, vegetables (cabbage) having water content in the range of 70-90 %. The green feeds are widely fed to rabbits, especially by small-scale producers. However, cost per unit of nutrients is too high for the commercial producers and the high water content renders them bulky, too low in energy for efficient production of meat or for lactating females.

2) Dry Roughages

The dry roughages, primarily hay (10-15 % water content) normally make up about 40-80 % of the diet for rabbits. While relatively low in energy value, the hays are economical sources of many nutrients, notably protein, some vitamins, mineral elements and furnish necessary amounts of fiber for a balanced diet. Legume hays are superior to grass hays in that they are generally more palatable, higher in protein and calcium concentration and are preferred by the rabbit. In most parts of the country Lucerne is generally used.

3) Concentrate Feeds

Those having low fiber content and relatively high digestibility and therefore high in nutritive value.

a. *High carbohydrate concentrates*:
   Those relatively high in more digestible carbohydrates but low in protein content. The cereal grains (maize, wheat, oats, etc.) and cereal byproducts (rice bran, wheat bran) are some of the examples for this group. These may be ground and fed as a meal or compressed into pellets.

b. *Protein supplements*:

   Those that are high in protein and are used to increase the level of protein in the total diet to the recommended level. Protein supplements of most interest to rabbit growers are derived from plants rather than the more expensive animal protein supplements. The most widely used are soybean meal, groundnut cake, sesame meal, linseed meal, etc.

   If cottonseed meal is used, it should be treated for removal of gossypol (cooking, lime processing, etc) and even then it should be limited to no more than about 5-7 % of the diet.

4) Miscellaneous Feeds

The group includes a wide variety of ingredients that are of little interest to commercial producers but important to backyard rabbit production. Table wastes (scraps) except meat, fat or spoiled foods are acceptable. In some places many weeds are used by small producers.

5) Feed Additives

Feed additives are materials that are administered to the animal to enhance the effectiveness and exert them in the gut or on the gut wall cells. There are clear theoretical reasons why these additives should be effective but, given the dynamic nature of the gut physiology, it is often very difficult to determine the effect in practice (e.g. Organic acids).

These include antibiotics, antioxidants, probiotics, prebiotics, nutraceuticals, enzymes, organic acids, hormones, emulsifying agents, flavors and pellet binders.

Antibiotics : They reduce the number of bacteria in the gut and increase the availability of nutrients to the animal. The widespread use of antibiotics has resulted in populations of bacteria that are resistant to many antibiotics and in several countries their use for growth promotion is redistricted or forbidden.

Probiotics : In contrast to the use of antibiotics as nutritional modifiers, which destroy bacteria, the inclusion of probiotics in foods is designed to encourage certain strains of bacteria in the gut at the expense of less desirable ones. A probiotic is defined as a live microbial food supplement that beneficially affects the host animal by improving the intestinal microbial balance.

Although the digestive tract of all animals is sterile at birth, contact with the mother and environment leads to the establishment of a various micro-flora. The beneficial micro-organisms produce enzymes which complement digestive ability of the host and their presence provides a barrier against invading pathogens. Digestive upsets are common at times of stress (e.g. weaning), and feeding with desirable bacteria, such as *Lactobacilli* (Probiotic), in these situations is preferable than using antibiotics, which destroy the desirable bacteria as well as the harmful species. The other bacterial species used as probiotics in mono-gastric animals include *Bacillus Subtilis and Streptococci.*

The desirable bacteria exert their effects in a number of ways;

a) Adhesion to the digestive tract wall to prevent colonization by pathogenic micro-organisms, e.g. E. coli.
b) Neutralisation of endotoxins produced by pathogenic bacteria which cause fluid loss.
c) Bacterial activity (lactic acid produced by Lactobacilli).
d) Prvention of amine synthesis (Coliform bacteria decarboxylate amino acids to produce amines, which irritates the gut concurrent with the incidence of diarrhea).
e) Enhance immune competence (Lactobacilli in young pigs elevated serum protein and WBC count).

Prebiotics: Prebiotics are defined as compounds, other than dietary nutrients, that modify the balance of the microfloral population by promoting the growth of beneficial bacteria and their by provide a healthier intestinal environment. Oligosaccharides occur naturally in foods. Soybean meal, rapeseed meal and legumes contain α-galacto-oligosaccharides (GOS), Cerials contain fructo-oligosaccharides (FOS), milk products have transgalacto-oligosaccharides (TOS), Yeast cell walls contain mannan-oligosaccharides (MOS). They are also produced commercially.

Enzymes: Not all compounds in foods are broken down by mammal's digestive enzymes, so some potential nutrients become unavailable. The chemical structure of plant cell wall (fibre), the limited time available for activity in certain parts of the gut and presence of anti-nutritive compounds (lignin/ tannin) in some foods hinder the release of nutrients. As a result of advances in biotechnology more effective enzyme preparations such as cellulose, xylanase, or pentasanase, â-gluatanase, proteases etc, can now be produced in large quantities and relatively inexpensively. Therefore, supplementation of the diet as a means of improving nutritive value is becoming common practice. Commercial enzymes are derived from a wide range of sources, particularly fungi such as Trichoderma longibrachiatum, Aspergillus viger and Humicola insolens. The enzyme should be selected on the basis of its efficiency of utilization of nutrients in mono-gastric animals, by releasing them for uptke from the small intestine rather than allowing fermentation in the hind gut, which results in products of lower value to the animal (e.g. Volatile fatty acids).

Certain feeds are of interest because of their detrimental effects. There is evidence that cottonseed meal may have cumulative toxic effect since rabbits are highly sensitive to gossypol in untreated cottonseed meal. Feeding large amounts of cabbage or rapeseed produces goiter in rabbits as well as in other species. Subabul has a toxic amino acid, mimosine. Subabul leaves should not be fed more than 10 % of total dry matter intake capacity of rabbit. If it is fed at higher levels hair loss occurs in rabbits.

Pelleted vs Non-Pelleted Feeds

Rabbits prefer pelleted diet to one in a meal form. Some individuals may refuse to consume a non-pelleted diet while others adjust to a meal diet and accept it satisfactorily, but during the adjustment period intake may be very low and feed spillage is excessive. Molasses or fat has to added to the meal diet to reduce dustiness and to increase palatability. Molasses is added at 3-5% in rabbit ration. Growth rate and efficiency of feed utilization were better with the pelleted diets.

Digestion and Metabolism of Nutrients

Rabbit is a non-ruminant herbivore. It is a hindgut fermenter. Digestive system is well adopted for the digestion of the large quantity of forage typical of herbivorous diet. Digestive system occupies a large proportion of body cavity and is nearly fully developed by 9th week of age. Caecum is major site of microbial fermentation. Microorganisms ferment the nutrients reaching the caecum. Most of the digesta (80% or more) is contained in the stomach and caecum.

Caecal Fermentation

In ruminant animals, volatile fatty acids (VFA) produced as end products of microbial fermentation in the rumen are absorbed across the rumen wall into the portal blood, and constitutes the primary absorbed energy source. On diets high in fiber, acetic acid is the major VFA produced in rumen, while on high concentrate diets, propionic acid production is increased. Butyric acid generally is produced only in small quantities of about 5-10% of total VFA. Butyrate is metabolized during absorption to ketone bodies and b- hydroxybutyrate in ruminants.

The VFA are also produced as end products of bacterial fermentation in rabbit's caecum. Use of VFA as a source of energy has been estimated to provide 30-40% of the maintenance energy requirements. It is demonstrated that VFA are a major energy source for metabolism of the hindgut tissues. Because of the use of energy requiring reactions in the absorption of electrolytes and other nutrients in rabbit colon, considerable energy is needed by hindgut tissues. Butyric acid is preferred fuel for hindgut metabolism followed by propionate and acetate.

The production and blood distribution of VFA are different in rabbits than in ruminants. In contrast to ruminants, in which butyric acid is produced in much lower quantity than acetate and propionate, in rabbit butyric acid is usually produced in considerably higher quantities than propionic acid. This is probably a reflection of the unusual microbial population of the rabbit caecum, dominated by *Bacteriodes* species, which often are butyrate producing organisms. *Lactobacilli,* which are propionate producers, are negligible in the rabbit caecum. Thus, the unusual pattern of VFA production in rabbit is probably a consequence of the unusual bacterial flora.

Caecotrophy

The rabbit produces two types of faecal pellets, one soft type and other hard type. Separation of digesta on the basis of particle size occurs in the hindgut. Peristaltic action rapidly moves large particles, primarily lignocellulose, through the colon and excretes them as hard faecal pellets. Antiperistaltic action moves small particles and solubles into the caecum, where they undergo fermentation. At intervals, the caecal contents are expelled as a "soft pellets" and consumed by rabbit directly from the anus. This process of eating faecal like soft pellets produced in the caecum is also called as caecotrophy or coprophagy or some times called pseudo-rumination. This type of feeding / reingestion provides microbial protein, vitamins and some quantities of VFA. The amino acids provided by microbial protein make only a minor contribution and diet must supply the additional amino acids.

Ration Formulation

In formulating diets for livestock, the objectives should be to design diets that meet the nutrient requirements of the animals and return the maximum economic benefit to the livestock owner. Diets that meet the requirements, with neither nutrient deficiencies nor marked excesses, are said to be balanced. The following types of information required to formulate diets: 1) The nutritional requirements of the animal 2) The nutrient composition of the feedstuffs available. 3) The availability of nutrients in the feedstuffs. 4) Non-nutrient characteristics of the feedstuffs, such as palatability and pelleting properties and 5) The cost of available feedstuffs.

Nutrient Requirements of Rabbits

The standard source of information on nutrient requirements of rabbit is the National Research Council (NRC,

1977 - Nutrient Requirements of Rabbits) publication from National Academy of Sciences, Washington, D.C., USA. The NRC (1977) requirements are given in Table-1. The French rabbit nutritionist Francois Lebas (1980) has established a set of nutrients requirements that is more closely oriented towards modern commercial rabbit production needs. The specifications are presented in Table- 2. In using nutrient requirement specifications, knowledge of what is truly important is needed. The nutrients of particular concern in rabbit diets are crude protein (CP), digestible energy (DE) and indigestible fiber (CF or ADF). As a general rule / thumb rule, the DE should be in the range of 2500-3000 kcal / kg of diet for rabbits. At lower levels, feed intake may be limited by gut capacity. At high energy density, hindgut dysfunction may occur with the development of enterotoxaemia or caecal impaction.

The level of indigestible fiber is important. Low levels of fiber are associated with ceacal-colonic hypomotility and development of diarrohea. Probably the best analytical measure of indigestible fiber is acid detergent fiber (ADF). A minimum level of 13% crude fiber or 21% ADF should be used for all classes of rabbits.

Least Cost Formulation using Computers

It is a routine practice to use computers in diet formulation. Programmes are available for ration formulation. Modern feed plants are computerized, so that the entire operation, including the weighing of ingredients is computer controlled. The advantages of using computer formulation are that many simultaneous equations can be solved, allowing the calculation of the least expensive combination of ingredients that will meet all the nutrient requirements. For each nutrient, a minimum and a maximum level can be specified. We might, for example, put 13 % CF as a minimum and 16 % as a maximum. The level of ingredients can be controlled. We may want to use at least 10 % wheat bran / rice bran but not more than 25 %. These minimum and maximum constraints can be specified.

Normally, least cost diet formulation is used, so that the least expensive ration is prepared.

Table 1 : Nutrient requirements of rabbits fed *ad libitum*[a,b].

Nutrients[c]	Growth	Maintenance	Gestation	Lactation
Energy and protein				
Digestible energy (kcal)	2500	2100	2500	2500
TDN (%)	65	55	58	70
Crude fiber (%)	10-12[d]	14[d]	10-12[d]	10-12[d]
Fat (%)	2[d]	2[d]	2[d]	2[d]
Crude protein(%)	16	12	15	17
Inorganic nutrients				
Calcium (%)	0.4	-[e]	0.45[d]	0.75[d]
Phosphorus (%)	0.22	-[e]	0.37[d]	0.5
Magnesium (mg)	300-400	300-400	300-400	300-400
Potassium (%)	0.6	0.6	0.6	0.6
Sodium (%)	0.2[d.f]	0.2[d.f]	0.2[d.f]	0.2[d.f]
Chlorine (%)	0.3[d.f]	0.3[d.f]	0.3[d.f]	0.3[d.f]
Copper (mg)	3	3	3	3
Iodine (mg)	0.2[d]	0.2[d]	0.2[d]	0.2[d]
Iron	-[e]	-[e]	-[e]	-[e]
Manganese (mg)	8.5[g]	2.5[g]	2.5[g]	2.5[g]
Zinc	-[e]	-[e]	-[e]	-[e]
Vitamins				
Vitamin A (IU)	580	-[e]	>1160	-[e]
Vitamin A as carotene (mg)	0.83[d.g]	-[h]	0.83d.[g]	-[h]
Vitamin D	-[i]	-[i]	-[i]	-[i]
Vitamin E (mg)	40[j]	-[e]	40[j]	40[j]
Vitamin K (mg)	-[k]	-[k]	0.2[d]	-[k]
Niacin (mg)	180	-[l]	-[l]	-[l]

Contd...

Pyridoxine (mg)	39	–[l]	–[l]	–[l]
Choline (gm)	1.2[d]	–[l]	–[l]	–[l]
Amino acids (%)				
Lysine	0.65	–[i]	–[i]	–[i]
Metheonine+cystine	0.6	–[i]	–[i]	–[i]
Arginine	0.6	–[i]	–[i]	–[i]
Histidine	0.3[d]	–[i]	–[i]	–[i]
Leucine	1.1[d]	–[i]	–[i]	–[i]
Isoleucine	0.6[d]	–[i]	–[i]	–[i]
Phenylalanine+ tyrosine	1.1[d]	–[i]	–[i]	–[i]
Threonine	0.6[d]	–[i]	–[i]	–[i]
Tryptophan	0.2[d]	–[i]	–[i]	–[i]
Valine	0.7[d]	–[i]	–[i]	–[i]
Glycine	–[e]	–[i]	–[i]	–[i]

[a] National Research Council (1977).

[b] Percentage or amount per kilogram of diet.

[c] Nutrients not listed indicate dietary need unknown or not demonstrated. TDN, total digestible nutrients.

[d] May not be minimum but known to be adequate.

[e] Quantitative requirement not determined, but dietary need demonstrated.

[f] May be met with 0.5% NaCl.

[g] Converted from amount per rabbit per day using an air dry feed intake of 60 gm/day for a 1 kg rabbit.

[h] Quantitative requirement not determined

[i] Probably required, amount unknown

[j] Estimated .

[k] Intestinal synthesis probably adequate.

[l] Dietary need unknown.

Table 2 : Nutrient requirement of rabbits recommended by Lebas(1980) [a]

Nutrient	Class of rabbit				
	Growing 4-12 weeks	**Lactation**	**Gestation**	**Maintenance**	**Does and litters fed one diet**
Crude protein (%)	15	18	18	13	17
Amino acids (%)					
Sulfur amino acids	0.5	0.6	-	-	0.55
Lysine	0.6	0.75	-	-	0.7
Arginine	0.9	0.8	-	-	0.9
Threonine	0.55	0.7	-	-	0.6
Tryptophan	0.18	0.22	-	-	0.2
Histidine	0.35	0.43	-	-	0.4
Isoleucine	0.60	0.70	-	-	0.65
Valine	0.70	0.85	-	-	0.8
Leucine	1.05	1.25	-	-	1.2
Phenylalanine	1.20	1.40	-	-	1.25
Crude fiber (%)	14	12	14	15-16	14
Indigestible fiber (%)	12	10	12	13	12
Digestible energy (kcal/kg)	2500	2700	2500	2200	2500
Metabolizable energy (kcal/kg)	2400	2600	2400	2120	2410
Fat (%)	3	5	3	3	3
Minerals					
Calcium (%)	0.5	1.1	0.8	0.6	1.1
Phosphorus (%)	0.3	0.8	0.5	0.4	0.8
Potassium (%)	0.8	0.9	0.9	-	0.9
Sodium (%)	0.4	0.4	0.4	-	0.4
Chlorine (%)	0.4	0.4	0.4	-	0.4
Magnesium (%)	0.03	0.04	0.04	-	0.04
Sulfur (%)	0.04	-	-	-	0.04
Cobalt (ppm)	1	1	-	-	1
Copper (ppm)	5	5	-	-	5
Zinc (ppm)	50	70	70	-	70

Contd...

Iron (ppm)	50	50	50	50	50
Manganese (ppm)	8.5	2.5	2.5	2.5	8.5
Iodine (ppm)	0.2	0.2	0.2	0.2	0.2
Vitamins					
Vitamin A (IU/kg)	6000	12000	12000	-	10000
Carotene (ppm)	0.83	0.83	0.83	-	0.83
Vitamin D (IU/kg)	900	900	900	-	900
Vitamin E (ppm)	50	50	50	50	50
Vitamin K (ppm)	0	2	2	0	2
Vitamin C (ppm)	0	0	0	0	0
Thiamine (ppm)	2	-	0	0	2
Riboflavin (ppm)	6	-	0	0	4
Pyridoxine (ppm)	40	-	0	0	2
Vitamin B_{12} (ppm)	0.01	0	0	0	-
Folic acid (ppm)	1	-	0	0	-
Pantothenic acid (ppm)		20	-	0	0 -

Simple Methods

Diets can be hand formulated, since the nutrient requirements of rabbit are simple. It is probably adequate in many cases simply to calculate the composition using a simple calculator.

Similar calculations could be performed to calculate energy (DE), calcium and phosphorus levels. This simple diet is in fact adequate for rabbit production. Improvements might include adding molasses for palatability and improved pellet quality and a small amount of fat for increasing ease of pelleting. For lactating does, addition of 1% dicalcium phosphate (DCP) and a vitamin premix would be advisable.

Consumption of feed by rabbit depends on the climatic condition. At ambient temperature of 30°C or above, the quantity of feed consumed reduces considerably. Any change in the feeding regimen should be gradual. If the feed is changed abruptly, the animal may refuse eating or it may lead to the development of digestive disturbances. Improper or inadequate nutrition of rabbits may lead to rough coat, poor growth, poor reproductive / productive performance and higher rate of morbidity and mortality.

Examples of complete feeds for rabbits

Nutritive value TDN (%)	CP (%)	Feedstuff	Level of Incorporation[1] (%) Grower diet	Lactation diet	Maintenance diets 1	2	3
60	15	Cowpea hay	30.0	35.0	50.0	50.0	50.0
80	10	Maize grain	28.0	40.0	10.0	10.0	-
65	14	DORB/ wheat bran	25.0	05.0	33.0	38.0	46.0
75	45	GNC(EXP)	15.0	18.0	05.5	-	02.0
		Mineral mixture	01.5	01.5	01.5	01.5	01.5
		Salt	00.5	00.5	00.5	00.5	00.5
			100.0	100.0	100.0	100.0	100.0

[1]On as feed basis

Feeding schedule for rabbits at IVRI, Izatnagar

Body weight (kg)	Concentrate feed(g)	Roughage[1] green (g)	Remarks
	Maintenance, Does and Bucks		
02.0	50	200	Equivalent amount of hay can be fed
02.5	60	250	"
03.0	70	300	"
	Pregnancy, Does and Bucks		
02.0	60	250	"
02.5	70	300	"
03.0	80	350	"
	Lactating , does with litter		
02.0	140	350	"
02.5	150	350	"
03.0	160	350	"
Growing			
	40-60	100-200	"

[1]Roughage sources like lucerne, berseem, cowpea or other grasses may be fed either as green or as good quality hay.

CHAPTER 8

Sanitation and Disease Control

Sanitation in the Rabbitry is the best disease-prevention system. Daily cleaning of cages or hutches, containers and surrounding is the easiest way to ensure sanitary condition and hence control disease. The most common diseases found in rabbits are cocciodosis. Pasteurella type pneumonia and enteritis. There are generally caused by lack of sanitation in the rabbitry. With strict sanitation practices such as cleaning all cages and water containers everyday, removal of manure, soiled bedding. Contaminated feed and collecting roughage from uncontaminated areas can drastically reduces chances of any disease in the rabbitry.

Isolate animals suspected of being diseased. Healthy stock exhibit bright eyes, glossy coat, firm flesh and vigour. Watch the droppings and take action at the first sight of any diarrhea(loose motions). It is usually safer to destroy a few animals than it is to treat them and risk spreading infections to healthy animals. This is especially true when some animals have respiratory troubles. One or more of the following signs are indications of a disease or abnormality, and each should be evaluated carefully. Failure to eat, abnormal breathing, eye and nasal discharge, loss of weight and unthriftiness, diarrhoea, listlessness and dullness, change in colour of mucus membrane, blood in excretions, skin lesions, loss of hair etc.

Some Peculiarities and Conditions in Rabbits

1) Coprophagy

Means the habit of eating faeces. The rabbits faeces, also known as droppings or faecal pellets, are round in shape. The rabbit produces two types of faecal pellets- the hard faeces and soft faeces. Hard faeces are mostly produced during the day time and soft faeces during the night. The rabbit has a peculiar habit of eating its soft faeces when it is not being observed. This habit is quite natural to the animal. Many people are not aware of this practice. Some who have observed it, believe that it indicates a nutritional deficiency and lack of quantity and quality of feed. It is, however, normal in rabbits.

2) Fur Eating Habbit

Rabbits that eat their own fur or bedding material or pull the fur from other rabbits, usually do so because the diet is inadequate in quality and quantity. A common cause is a diet low in fibre or bulk. Some times the protein content of the diet may be too low. Adding more Jowar, groundnut cake, soyabean or any other legume may correct the deficiency. Providing good quality grass or feeding fresh leguminous feed or root crops as a supplement to the home- mix or concentrate diet also helps to correct an abnormal appetite.

While cleaning themselves by licking their coats or when eating fur from other animals, rabbits swallow some fur which is not digested. The only noticeable result may be droppings fastened together by fur fibres. However, if the rabbit swallows a lot of fur, it may collect in the stomach and form a "fur block" which interferes with digestion. If it becomes large enough, it blocks the alimentary tract and the animals starve. The most satisfactory method is to try to prevent fur eating by providing adequate roughage and protein in their diets. A block of wood sooked in salt solution for three days and hung in the cage for the rabbits to chew may be used to reduce fur chewing.

3) Gnawing Wooden Parts of the Hutch

The rabbits have sharp teeth and gnawing wood is natural for the rabbit. However, in excess, it may indicate insufficient amounts of salt in the diet. Rabbits that have access to good quality grass and those receiving legumes and rootcrops are less likely to gnaw on their hutches.

4) Malocclusion or Buck Teeth

The incisors (front teeth) grow long so that the mouth cannot close properly. The upper teeth grow downwards and curl backwards, and the lower grow upwards and protrude. Animals can not eat prperly. Do not use such animals for breeding. Cull rabbits showing buck teeth.

5) Molting or Shedding of Hair

Sometimes the shedding of hair begins when the rabbit is about 2 months old and lasts until it is four to six months of age. Heavy feeding of the young tends to cause the molt at an earlier age. In addition, there is the annual molt with mature rabbits. This molt varies in time of appearance in different geographical areas.

Rabbits may also molt in disease, going off-feed, the sudden occurrence of unseasonably high temperature or other stresses. Evidence of molting is the accumulation of loose hair in the cages, and the droppings being bound together by embedded fibres.

Shedding first appears on the sides of the hips and thighs, fallowed by appearance on the back, and then increasing in areas over the sides. In places where the hair shed, new hair (coat) also grows.

6) Sore Hocks

There is loss of hair on the foot pads and a big sore on the hocks. It is a disease. The rabbit sits with more weight on the front feet normal, and will also walk with an unusual motion. An inspection of the feet quickly reveals sore hocks. A dry, ulcerated and crusty scab appears on the hocks of hind feet and sometimes front feet. There may be loss of appetite and weight because of pain. Small lesions may be helped by placing resting boards or mats in the cage. Advanced cases are best culled. Applying iodine ointment may be helpful.

Sore hocks may affect the vitality of does and bucks, especially at the time of mating. Breeding of the rabbits with this condition is not advisable.

7) Canker

It is a condition of the ear. Brown scaly crusts are found at the base of the inner ear. This is due to infestation by mites (small parasites that burrow into the skin) through feeds or unsanitary cages. Remove scales and crusts from inner ear and swab with a mixture of 1 teaspoonful of edible oil and 3 drops of iodine.

8) Coccidiosis

It is a disease that must be considered when enteritis (diarrhoea) is observed. The faeces will have blood. Whenever enteritis is observed in your rabbits, collect faeces and take it to a Veterinary clinic for examination. Medicate the water with sulfaqinoxaline (Sulmet, Amprolium, Bifuran) and follow the directions of its use. All the rabbits should be treated simultaneously and not only to those when show symptoms.

CHAPTER 9

Slaughtering and Skinning of Rabbits, Shearing and Preparation of Products

Important by product from Rabbit are meat, wool, fur and skin. It can be made highly profitable if products are properly processed and marketed. Many rabbit products are wasted because people do have not enough knowledge about processing.

Slaughtering: Know when and how.

Skinning

Familiar with how to remove skin and carcass preparation. How the skin has to be processed and how the products which can be obtained from skin.

Manure collection and preservation.

Slaughtering : It depends on demand, type of animal where meat is consumed more, it can be achieved in 2 methods.

Techniques

Before slaughtering, animal should not be frightened for humanitarian purpose. Adrenaline affects the quality of meat.

1. The animal is killed by 2 methods:
   a. By dislocating neck by simple quick jerking (Twisting).This can be done by holding rabbit upside down. One hand holds the animal (hind legs) and with other hand he makes a simple jerk of neck by sharp downward pressure and simultaneously twist neck round and backward. This method of killing is employed for fattened rabbit which is too heavy.
   b. By administrating sharp blow to back of head with the help of stick or iron bar. This can be done placing the animal on the table and with the help of stick in iron bar hit behind the skull. This is mainly used for aged and old animals.
2. a. Immediately after killing, hang the animal by its hind legs on the which is provided with sharp pointed hooks. These hooks are permanent by fixed to the wall.
   b. The other method is by use of twine or rope. Tie one end of rope to the wall hooks or rods on ceilings. The rope should be strong enough to withstand body weight of animal. Another method is by using shackles.
3. Then cut off head and front leg after hanging (cut at elbow joint).
4. Make a circular cut at hock joint of hind limb. Then try to make the cut towards the bdominal region till it reaches the hock joint of other hind at which point make one more circular cut of the hock joint.
5. Then cut off tail.
6. From this cut, pull the skin downward. Continue peeling it until you have pulled it completely from the body.
7. Make a slit near the belly portion of animal. Take care not to cut the intestine at the time of incision.

8. Take out all the visceral organs expect liver, kidney and heart.
9. Wash the carcass in running water.
10. Store the carcass under refrigeration temperature.

Slaughtering a rabbit

You will find it easier to slaughter rabbits than any other farm animals. With experience, you can do whole jon in two or three minutes. Follow these steps :

1. Kill the rabbit by dislocating the neck with a quick jerk.

2. Hang by its hindlegs on two sharply pointed hooks permanently attached to the wall

   A second approach is to use a piece of rope or twine. Be sure that the rop is strong enough to bear the strain when pulling off skin.

   Other approach include the use of shackles as can be seen in No. 3

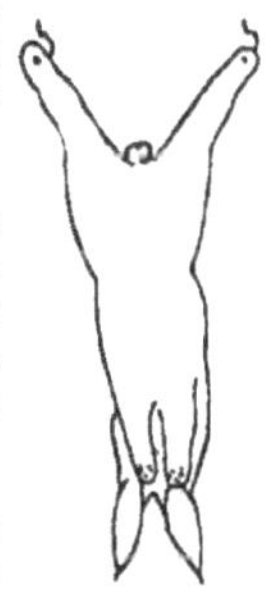

3.

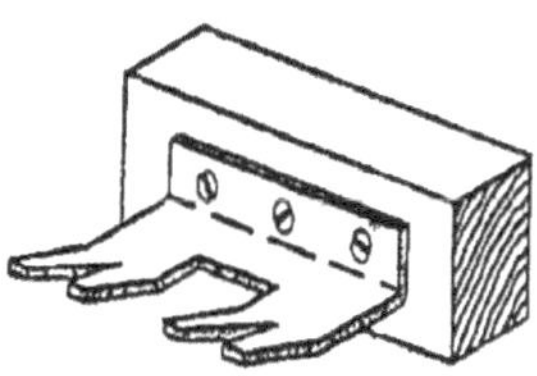

A shackle made from steel plate.

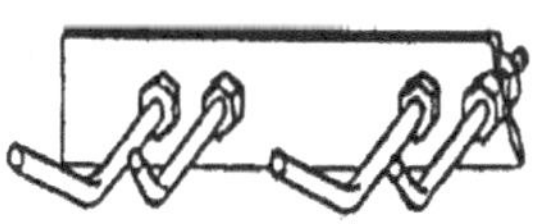

A shackle made using bolts

4. Cut off the head and front feet

5. Cut the skin around both hocks. Cut open the skin on the inside of one leg at the hock joint, working your way towards the tail and up the other leg.

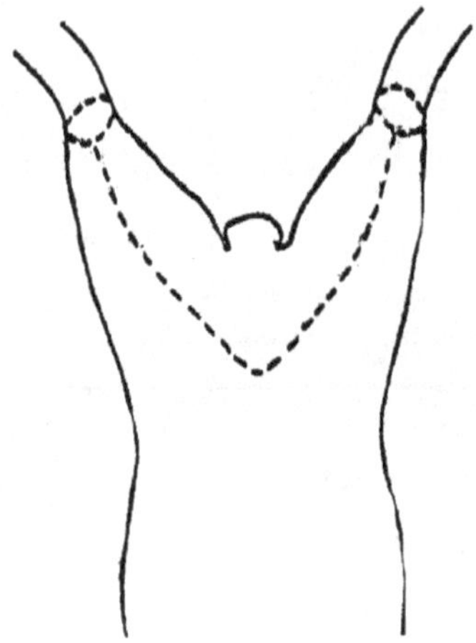

6. Cut the tail off. Start to peel off the skin off both hindlegs. Continue peeling it downward until you have pulled the skin completely off the body.

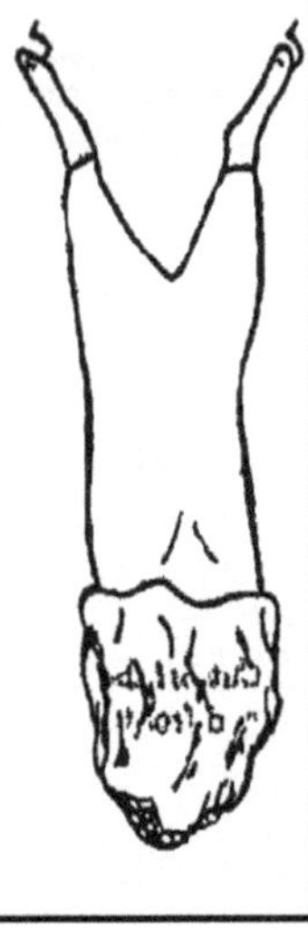

7. Slit open the belly. Take care not to cut the guts. To finish remove everything inside except the kidneys, liver and heart. Wash well with clean water. Cut up the carcass in to portions or cook whole.

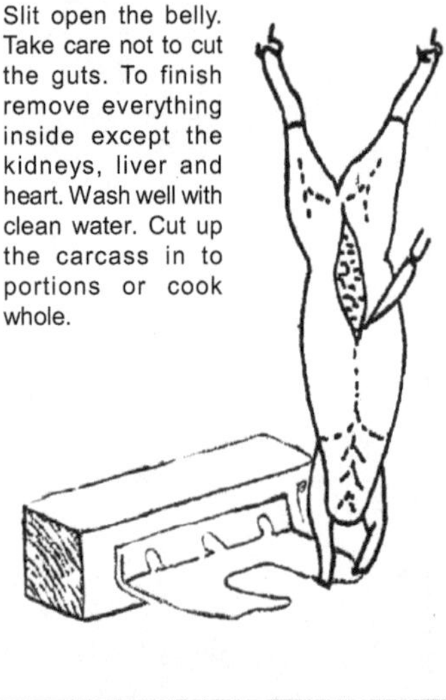

Skinning of the Animal

For skinning, knife or scissors should be sharp and clean.

1. For skinning, animal should be killed and hanged. The tall and front feet is cut and even head.

2. Make a circular cut at hock joint and while running incision join it to other hock joint.
3. Pull the skin downward over the body surface peeling adhesive tissue and fascia.
4. Pull the skin from neck and fore limbs. At the time of peeling, care should be taken to avoid blood stains.
5. Immediately after peeling put the skin on stretcher which has metallic board with pins for properly fixing nails on all corners. This avoids wrinkles.

Preparation of Carcass

1. The carcass is handled from the fixed hooks after skinning.
2. Make an incision from anus to rib region. Care should be taken not to tear Intestine and other viscera.
3. Remove stomach, intestine, bladder but keep heart, liver and kidney intact.
4. Take carcass from hanging position and cut off hind legs.
5. Clean the carcass with water to remove fur, dirt and blood stains which are present at the time of slaughtering.
6. Cloth should not be used for cleaning as it causes contamination.
7. The carcass can be cut or preserved as such at room temperature for 24 hrs and under refrigeration upto 1 week according to requirement.

Treatment of Skin

Only if skin is properly processed, it will have a good market.

1. After skinning, skin should be dried. This is called curing. For this the skin should be put on stretcher with nails to avoid wrinkles. These should be protected from insects and pests. This can be done by applying /sprinkling cooking salt.

2. Process the skin to improve market value and this is called tanning: tanning involves two methods.
   a. Alum tanning
   b. Chrome tanning

In alum tanning common chemicals are used and it is a simple method.

For chrome tanning, chemical like conc. Sulphuric acid which cause burning is used, so not much useful.

Alum Tanning

The skin is made into a leather form which can be used for making usable products like caps , fur coats , toys and dolls.

Steps

1. Remove the skin from stretcher.
2. Skin soft portion is placed on the wooden board and remove excess fat and fascia with the use of blunt knife.
3. Rinse the skin in soap water containing 5-10 g of washing soda or Borax in 1 liter of water. Squeeze the excess water from skin.
4. Preparation of chemical - Dissolve ½ kg of ammonium aluminium SO4 in 5 liters of water or ½ kg of potassium aluminium sulphate in 5 liters of water - solution 1. 150g of washing soda is dissolved in 3 liters of water - solution 2.
5. Mix solution 1 & 2 thoroughly preventing formation of lumps.
6. Apply the layer of the paste on soft layer of skin fixed on wooden board and keep it for 24 hrs.
7. After 24 hrs. apply one layer of paste.
8. After it is dried, sprinkle some quantity of saw dust and apply little quantity of vegetable oil which gives shiningness to skin.

This type of processed skin can be preserved for months together.

Rabbit Manure

Has value as good organic fertilizer .It can be collected by providing mesh flooring (1/2 ″ x ½ size) and facilitate of manure.

Preservation of Manure

This is called composting

1. Use of earth worm.
2. Making use of vegetable matter and vegetable leaves manure on these

   On manure layer , one more layer of vegetable leaves. Keep adding water to avoid spoilage and smell of manure.
3. Also on top , put a layer of soil or sand . By this method, manure can be stored for many days.
4. In dry form - to get maximum utility compositing is done.

Rabbit manure: The rabbit manure is very much useful as organic manure in agricultural field and for fruit or horticultural plants or crops. The main advantage of rabbit manure is that, it can be applied as raw animal manure directly from rabbit sheds to vegetables or agricultural plots or fruit plants. It does not cause any harm to the crop or plant as done by raw cattle dung. The rabbit manure is richer in nitrogen, phosphorous and potassium compared to cattle manure.

The quantity of manure produced per year by rabbits depends on size, age and breed of rabbits and type of feed they consume. However on an average an adult rabbit produces about 100 - 150 kg of manure per year.

Rabbit manure may be applied to the soil for feed crops at 5-10 tones/hectare. In vegetable gardens it can be used ar 10 kg/ meter square, preferably in 2-3 doses. For fruit bearing plant 10-13 kg may be applied.

Table Comparative values of manure of rabbits and other animals

Species	Nitrogen (%)	Phosphorous (%)	Potassium (%)
Rabbit	3.7	1.3	3.5
Cattle	2.9	0.7	2.1
Sheep	3.2	0.9	2.8
Poultry	4.1	1.6	1.0

Shearing

Angora rabbit can be sheared every 3 months and at every shearing 250 g of wool. On an average in 1 year, 1 kg of wool can be obtained.

Good Angora breed yield wool of 2-21/2″ on average. On record, 8″ of wool is also obtained.

Classification of Wool

Based on length and location.

1. Jares -10-12 cm in length. Present on outer surface.
2. Barts
3. Duvets

Jare – gives protection to animal. So called 'Guarding hair'

Barts – 8 cm in length. In distribution, they are 4 times than jares.

Duvets-5 cm in length. These are finest wool.

Shearing is done just prior to mating. During mating , wool may get matted.

Methods of shearing / removal of wool:

1. Plucking the wool
2. Clipping
3. Shearing
4. Administering chemicals

Plucking

With the help of hand, wool is plucked from body. It causes damage to animal, is practiced in China and France . Full length wool is obtained.

Clipping

Done with clipper or barber's scissors. Shearing and clipping, equipment required are table with belts. Belt is to restrain the animal, pair of scissor or shearer, measuring rod ruler to measure length of wool. Start clipping or shearing from the neck of animal to go backward to lateral side extending upto tail. At the time of shearing or clipping, care should be taken not to make skin cuts .For female, teat should be taken care of, pregnant animal is not sheared.

1. To avoid damage to young ones.
2. Require wool to make nest.

Chemical Removal of Wool

When mimosin is fed, failing of hair results .In sheep, in western country excess of subabul (Leucaena leucocephala) which has mimosin .This causes deflects. This is temporary .For broiler rabbit , subabul can be fed at limited quantity (10 - 20 % of the total nutrient requirement, 28% protein - leguminous) and good growth is obtained and good marketing weight is obtained.

Processing of Rabbit Skin

Rabbit keeping is undertaken mainly for the production of meat. The skins and the manure are by-products. Many rabbit products are wasted because they are not properly treated. If the rabbit keeping are knows what to do with these products, he will find rabbit keeping much more profitable.

If the main reason for keeping the rabbits is for meat, then the skins are a by-product. They are unlikely to be of high quality, especially those from young rabbit under six months of age. There may or may not be a market for them locally. The breeder may wish to keep a few for use in the home. If the skins are to be sold or used, the first step is to dry or cure them, then to treat them chemically.

Drying the skins known as curing and it is carried out as soon as possible after skinning the animal. The skin is slipped over a wire stretcher with the fur to the inside. It is then hung up to dry in a suitable place, shaded from the sun and protected from pests. Any pieces of fat should be scrapped off with a blunt knife. The curing should only take a few days after which the skins should be sold as soon as possible.

Rabbit skin forms one of the major commercial products from the rabbits. In western countries up to 40 per cent of the income on rabbit rearing is through the skin. But in our country the market is not developed well and hence many times the skins are destroyed without any benefit for the farmer. Rabbit fur production is not comparable with other species of livestock's. Mink, which tops the list of species bred essentially for its fur, supplies a world total of about 25 million to 35 million pelts a year whereas, rabbit pelts are estimated at one billion. In France alone annual rabbit skin production tops 70 million.

Few skins are now retrieved from slaughter houses: they are simply thrown away.

Those that are used fall into three categories: fur pelts for dressing, pelts for shorn hair (hair removed from skin) and skins for use as fertilizer. The most important aspect of rabbit skin processing is to collect best skins of rabbits. Skins of rabbits with skin diseases, alopecia etc need to be avoided. Too much older animals are also not suitable as their skin will be thick and may require more time for processing and bulk processing it will create problems.

The animal should be slaughtered scientifically and its skin be removed without damage, blood or dirt immediately. Rabbit skins if kept for more than half an hour will start decaying and this will immediately result in loss of hairs and thus the market. If skins are kept together it can cause damage to the skins through enzyme actions and will result in of revenue.

Skins are Sorted in to Three Grades

1) *Pelts for dressing*: (the term "dressing" instead of "tanning" is used for fur). These are the best skins, with regular shape, intact, homogeneous, and dense, a well-formed coat, a flawless skin. Their price may be 20 times that of unsorted quality skins.
2) *Pelts for shorn hair*: These usually lack the proper shape or are not homogeneous enough for fur products. The hair, however, is sufficiently long and healthy. It is therefore machine shorn and used for textiles or felting (although the hat trade is declining industry) or fertilizer. This technology allows much of the pelt to be recycled.
3) *Waste*: Unusable except for fertilizer (the hair is gnawed, cut, soiled, sweaty, parasite ridden). Such skins push up the costs of lobour, processing and transport.

In France, one of the foremost rabbit producing countries, the proportion of pelts suitable for dressing is less than half of those collected. The figure differs from one author to another, which is not surprising in view of the difficulty in getting exact data on this product.

Rabbit skin can be stored before processing in two forms *viz,* dry and salted forms. Preparation of dry rabbit skin is to be done between 18^{0}C and 22^{0}C. Exposure to direct sunlight will cause damages to scratch out the excess meat and fat from the inside surface of the skin. Care should be taken to see that time of drying skins are not touching each other. Before drying skin should be put in a stretcher so that wrinkles or folds are not produced in the skin. Wooden or metal stretchers can be used for the purpose. The salted skin is produced by pasting salt on the inside part of the skin. Disadvantage is that it has to be stored under refrigeration, other wise it will be damaged.

Alum-Salt method

One of the easy methods to process the dry skin is alum - salt method. The different steps involved as follows.

- Cut the dry skin in to two through the centre
- Rinse it in cold water many times. The water should be changed each time.
- Remove the meat and fat scrapping with a blunt knife.
- Dissolve 5 - 10 Borax in one liter water and rinse the skin solution many times. Press out the excess water.
- Prepare the following solutions

a. Dissolve 500 grams Ammonium aluminium sulphate or Potassium aluminium sulphate in 5 liters of water.

b. Dissolve 150 grams washing soda and 300 grams sodium chloride in 3 liters of water.

- Mix the two solutions together.
- Make a paste by mixing flour with the mixed solution
- Fix the skin with the hairless surface upwards on a wooden board with nails. Apply the paste on next day and reapply fresh paste once more. Leave it for few days.
- Remove the paste by scrapping with a blunt knife.
- Wash in Luke borax solution (5 - 10 grams/liter).
- Allow the skin to dry. Simultaneously stretch the skin side ways to make it soft.
- Rub the skin with wood dust to make it shining.
- Application of glycerin or other solutions used to soften the skin is also advised.

Alcohol – Turpentine Method

Most easy to process the dry skin is alcohol - turpentine method. The different steps involved as follows.

- Cut the dry skin into two through the centre
- Rinse it in cold water many times. The water should be changed each time
- Remove the meat and fat scrapping with a blunt knife.
- Prepare alcohol - turpentine mixture by mixing 500 ml each of alcohol and turpentine.
- Put the wet skin inside the solution and it stir it every day.
- Take the skin out after 7 - 10 days.
- Wash the skin in mild detergent solution to remove the dirt and oil.
- Wash the skin water and press it to remove the excess water.
- Allow the skin to dry. Simultaneously stretch the skin side ways to make it soft
- Rub the skin with wood dust to make it shining.

Salt – Acid Method

One of the other easy methods to process the dry skin is alum-salt method. The different steps involved as follows.

- Cut the dry skin into through the centre.
- Rinse it in cold water many times. The water should be changed each time.
- Remove the meat and fat scrapping with a blunt knife.
- Dissolve 5 - 10 Borax in one liter water and rinse the skin solution many times. Press out the excess water.
- Dissolve 500 gram salt in one liter water.
- Pour 500 ml concentrated sulphuric acid to salt water. This step should be done very carefully as the acid is highly caustic.
- Prepared solution can be stored in glass or earthen pots.

- After cooling the solution to room temperature, put the rabbits skins prepared into the solution. Care should be taken to see that the skins are immersed fully.
- Keep the skins in side the solution for two to three days with occasional stirring.
- Remove the skins from the solution and wash it in fresh water.
- Put the skins in Borax solution (5 - 10 gram/liter) and mix it gently.
- Take out the skin and press it to remove the solution.
- Allow the skin to dry. Simultaneously stretch the skin side ways to make it soft.
- Rub the skin with wood dust to make it shining.
- Application of glycerin or other solutions used to soften the skin is also advised.

Chrome Method

Compared to the other methods chrome method yield the durable products. The different steps involved are as follows;

- Cut the dry skin into through the centre.
- Dissolve washing soda and Noigene at the level of 0.2 per cent in luke warm water and put the skin inside the solution for one hour.
- Take out the skin and press it to remove the water.
- Fix the skin with the hairless surface upwards on a wooden board with nails.
- Remove the meat and fat scrapping with a blunt knife.
- Dissolve 50 gram salt in one liter water and put the skin inside it for ten minutes.
- Put the skin in 10 per cent sulphuric acid or formic acid solution for one hour with constant stirring.
- Then allow the skin to remain in the solution for one day.

- Dissolve 20 grams potassium dichromate in half liter warm water (80 to 90 per cent).
- After cooling the solution to room temperature, pour the chrome solution with acetate and sodium carbonate and pour it to the previous solution with constant stirring.
- If the processing is perfect, the processed skin should not shrink in hot water. To test this cut a small piece of skin and put it in water at 60°C and examine after one minute.
- Prepare lipoderm liquor II by mixing one part lipoderm liquor and three parts warm water at 70°C. Paint the solution on the hairless parts of the skin.
- Rub the skin with dust to make it shining.

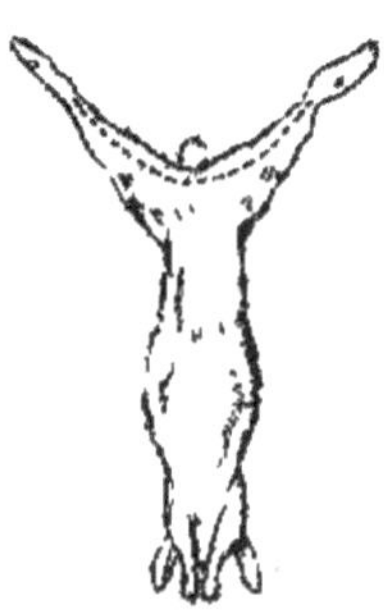

Figure 1. Skin cut between the thighs

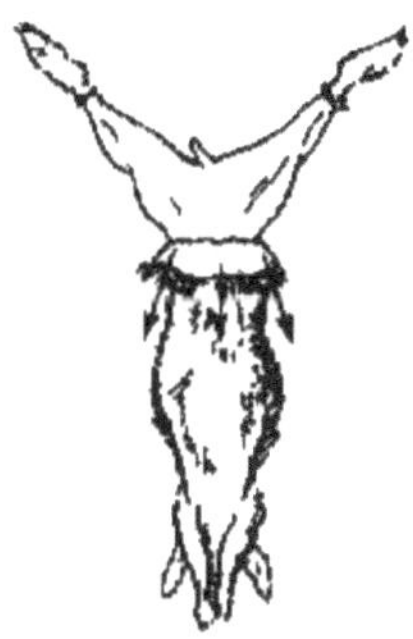

Figure 2. Skin pulled off the hind legs

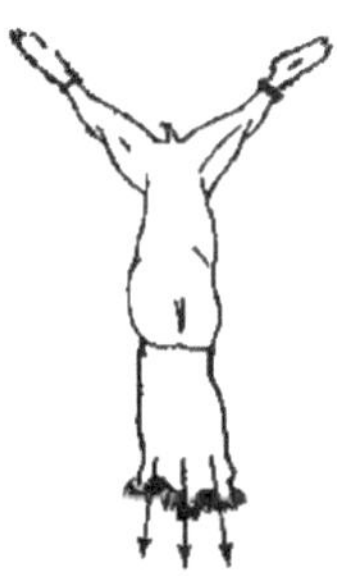

Figure 3 Skin pulled to bare the trunk and then the forelegs but not eviscerated

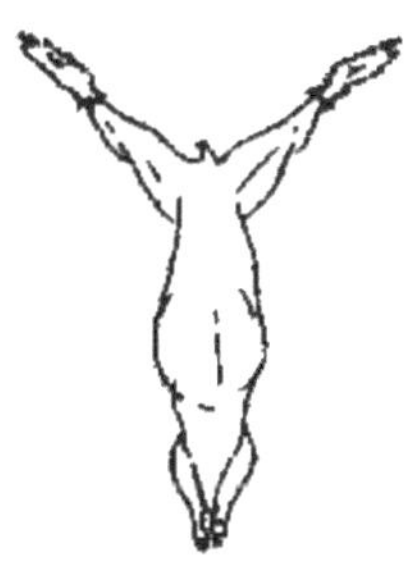

Figure 4 Carcass skinned

Figure 5. Carcass opened, white viscera (Intestinal tract and lungs) removed

Figure 6 Correct way to dry rabbit pelts (b)

Figure 7 Correct way to dry rabbit pelts (a)

CHAPTER 10

Angora Rabbit Wool Production

Food, furs, research, hobbies, pets - these are the major uses of rabbits around the world. It should be noted that rabbit production for meat and wool has become increasingly important in the last few years.

Angora wool rabbits are a commercial rabbit which is raised for its fine wool and meat as well, but primarily raised for wool production. Cooler climate is most conducive to raising this breed of rabbit. Angora rabbit does not differ from other breeds of rabbit except for its unusual length of valuable fur(wool). This modifies its raising compared to other rabbits.

The term 'Angora' qualifies those mammals which are covered with fur or wool composed of very soft hair of more than normal length - double or more than double the normal length. This characteristic is relatively found in some of the domestic species of cat, goat (mohair), rabbit, dogs and sheep. This is not found in wild animals except in animals of polar region.

It seems that the Angora rabbit was first sighted in England at the beginning of 18th Century Mortimer in 1708; then termed as "White Shock Turkey Rabbit".

The world production of Angora rabbit wool is approximately 2,000 tones per annum. The chief producers are the People's Republic of China (over 1,500 tones), as well as France. Czeckoslovakia , Argentina and Brazil.

DLG information bulletin 3 (1984) reports that since the beginning of research station studies on Angora rabbit wool production performance (ALP) in 1934 in Germany, the annual yield has gone up from mean weight of 292 g to over 1000g (males 874g. and females 1159g.) in 1970. Owing to the fact that higher wool production is accompanied by greater difficulties in thermoregulation and an apparent antagonism between the quantity of wool produced and reproduction performance, top performances of wool yields have obviously reached their physiological limits, at least as far as purebred animals are concerned.

There is an upward trend due to the fact that advances in spinning methods (upto 180 meters/g wool) have extended the use of Angora breeds.

Selection of Breeds

The two most popular varieties of angora are the English Angora and the French Angora. They have been created by a long line of selected rabbits having the type of wool desired by the wool markets. Normal fur is the fur all rabbits have when they are not rex, Satin or Wool, Angora is usually white and albino, but some colored varieties are also available. The wool of English Angora is silky in nature unlike the French strain which has coarse wool. The French Angoras are heavier than the English ones. The adult body weighs vary from 2.5 kg to

3.5 kg and the annual wool yield ranges from 400 g to 1000 g. Apart from these two strains, the German breed of Angora rabbit ranges amongst the most efficient Angora rabbit populations of the world. This has led to a considerable increase of the breeding stock exports during the last few years.

Normally the much development of Angora isd of white color and long staple for uniform dying and processing. In various countries different criteria have been adopted for rearing of Angora rabbits as such selection and development of breeds have been done accordingly. The following strains or breeds of Angora rabbits are being reared by the Indian farmers at present.

a. *Crossbred Angora*: Origin (cross of 2 or 3 breed) India, well adopted to Indian conditions, wool yield 500 - 600 g / year, wool white medium fine sound hair - 2-4%, adult body weight - 3-4 kg.
b. *German Angora*: Origin is Germany; wool yields 700-1000 kg/year, white fine quality. 2-4 % sound hair, adult weight is 3-4 kg.
c. *British Angora*: Origin UK, wool yield 400-600 g/year, white lustrous fine quality, sound hair 2-4 %. Adult weight is 2.5 - 4.5 kg
d. *Russian Angora*: Origin Russia, wool yield 300-400 g/year, medium fine quality wool, 10-20 % sound hair, adult body weight is 3.5 - 5.5 kg.

English Angora

Wool should be as dense as possible over entire body including belly, back, sides, chest and rump. It should be wavy and silky and from 21/2 to 2 ½ inches in length and as uniform and even as possible. Soft matted wool, unevenness in length and wool that is thin and lifeless should be avoided. The side profile should be slightly oval from the collar back and any breaks in the outline caused from thin or broken wool should be avoided. The white wool should be clean and free from stain.

The black wool should also be even, clear and deep. The fawn wool should be of good depth, even and clear. Density should be checked by both feeling and blowing into the wool. Thin wool and bare spots should be avoided. Sixty five points are allowed for colour, length, texture and density of wool.

French Angora

Wool is coarser than English Angora with a fair amount of guard hair. Length is from 2 ½ to 3 ½ inches and as dense as possible showing life. Colours should be identical to the English Angora. Extreme coarse or soft wool should be avoided. Wool should be in prime condition and fall free. Body should have an oval shape. Colour, length and density of wool over entire body are important. 65 points are allowed on colour, length density and texture of wool.

The American Rabbit breeders association, however, make English and French types synonymous. It is different to distinguish the English Angora rabbits from the French when they are cross-bred. The wool fiber of the French is shorter and coarser than of English , but the wool yield is greater.

Cross -breeding Angoras with other breeds has proven beneficial both for better meat and fur production. The commercial Angora weighs at least 8 pounds and is being bred more and more to improve its meat quality.

Texture, Density, Balance and Condition

According to the American Rabbit Breeders Association (1976), the commercial standard spreads the points in three divisions with the most important texture with 40 points, density 30 points, balance and condition 30 points.

Texture is visible but touch is the only way to determine it conclusively, and by comparison the judge makes the decision as to which is the best texture. The fly back explains itself and without fly back there is not texture of quality.

Density is that quality of thickness required to make full bodied fur piece and can be determined at the same time the fly back is checked. Immediately after the hand passes, When stroked from rump towards head, if the skin is too much in evidence, it denotes a lack of density. No amount of undercoat is sufficient to give density without the thickness of guard hair. So the guard hair is a must for density. These qualities of texture and density should cover as much of the rabbits body as possible, even to the belly. The more of the fur that can be used the more valuable is the pelt. Balance calls for the coat to be uniform in length and density over the entire pelt as much as possible and for a uniformity in the difference in length of the guard hair and the under coat.

Condition requires that the fur should be prime, not dead or in moult and that the fur be clean. The prime rabbit is clearly visible on close observation.

Table 1 : Points to determine the value of Angora rabbit (Angora rabbit– Book of France)

Conformation of body - Cylindrical	5
Head and ears (ears V shaped)	5
Development of body (3.75 - 4.25 kg)	10
Output of wool	30
Homogeneity and density of wool	10

Wool Growth

The better types of Angora will produce from 2 to 2½ inches of wool in approximately 11 to 12 weeks, or an annual growth of 8 to 10 inches. This represents 12 to 16 ounces of wool a year from a mature Angora that is not nursing young. This wool is valued for its softness, warmth and strength when properly balanced rations are fed; 50 kgs of feed is required on the average to produce 500 g of wool.

Retarded Wool Growth

Areas where the growth of the wool is retarded become evidence following removal of the baby coat at weaning age, and the patches that appear on the thighs of the rabbit may escape notice, although they may be more or less persistent and may remain for 8 to 10 they do not affect the wool yield materially because they cover small surfaces. There is considerable variation in the extent of the areas of retarded wool growth on other parts of the body. Most cases affect the coat only slightly and do not produce persistent spots but will cause areas of uneven growth that may escape detection unless one makes a special effect to locate the patches. It is not the result of a nutritional deficiency.

It is absolutely essential that records be kept of the quality and yields of wool, for these records are indispensable when matings for improving the herd are planned and when rabbits are culled.

The Angora rabbit has 3 types of hair. For external protection:

1) Jarres, 10 to 12 cm (guading hair).
2) Bards, 8 cm (the bards are 4 times more than the Jarres) for internal thermic isolation .
3) Duvets, 7 cm (they are the finest).

Jarrers have hair follicles. Bards are very soft and fine and do not give rigidity to the fur. Barbs and Jarres offer Protection to the animal. The body of the rabbit is devoid of sweat and sebaceous glands. Therefore, the fur remains clean, besides the animals have the habit of licking them selves clean.

Grooming and Shearing

Wool is harvested prior to breeding to prevent mauling and soiling the wool. There are 3 methods for removal, wool plucking, shearing and clipping. Each method has its Proponents. The annual yield will not be influenced by the method used. Some producers pluck their animals monthly and some at intervals beyond 3 months. Plucking is mostly practiced in China and France, and the operation takes 20 to 30 minutes. By plucking, a fur that is well structured and uniform is obtained with the hair morphologically intact i.e., the heads of bards and jarres are intact which is not the case with the use of shears or scissors. This plucked wool brings the highest price on the market but it is necessary to handle the rabbit three times in order to remove the entire coat.

The following equipment is needed for grooming and shearing:

- A table, waist-high, with a 12″ X 24″ top covered with a carpet or a feed sack to keep the rabbit from slipping and equipped with castors to allow easy turning. A table equipped with straps or cords for restraining the rabbit is advantageous.
- A Pair of barber's scissors or shears.
- A hair brush with single steel bristles set in rubber for brushing and removing foreign material from wool.
- Containers for storing wool.

Grooming

Commercial woolers require little, if any grooming between shearing provided they are properly cared for and sheared every 10 to 12 weeks. If the coat is allowed to grow for a longer period, the fibres may become webbed, tangled or matted.

For grooming, the rabbit are placed on the table. The wool is parted down the middle of the back and one side of it is brushed stroking downwards. As the end of the wool is reached, it should be brushed upward and outward to remove all foreign material, another part in the wool about half an inch farther down the side is made and the operation is repeated until the job is completed. Then the other side is groomed the same way.

For grooming the head, front legs and belly, the rabbit is placed on its back preferably in the operators lap. The hind quarters are gently held but firmly between the knees. Small areas of wool are separated and groomed the way the sides are done.

For grooming the hind legs, the rabbit is best placed on its back in the operators lap. The head and front feet are held under the arm and the left hand is used to old the rabbits hind legs.

Shearing

Before shearing, all stained ends of the wool are cut off. The back of the scissors or the shear is placed against the rabbits body and care is taken to prevent cutoff of the skin. The operation should begin at the rump and a strip of about an inch wide is sheared up to the neck. This operation is repeated until all the wool is removed from one side. Then the rabbit is turned around and shearing operation is accomplished on the other side, starting at the neck and shearing, towards the rump. For shearing the head, front legs, belly and hind legs the rabbit is restrained as for grooming. Small areas of wool are separated and sheared the way as done on the sides. Care should be taken not to injure the doe's teats. The wool should never be sheared from the belly of a pregnant doe. After shearing, the rabbit should be lightly brushed to straighten out the wool fibers and prevent the formation of mats.

Grading, Preparation and Marketing Wool

A container for each grade of wool is graded as sheared allowing the usual commercial grades.

Plucked Wool

Super - 0.825 m or longer.
No.1 - 0.75 m or longer.
No.2 - 0.05 m or longer.

- Pure white absolutely clean wool, free of all mats and foreign matter, staple length 2¼ to 3 inches.
- Pure white, absolutely clean wool, free of all mats and foreign matter, staple length 1½ inches.
- Pure white, absolutely clean wool free of all mats and foreign matter, staple length 1 to 1½ inches. Baby wool will grade as No.3 because the fibers lack tensile strength. "Shorts"
  - Pure white and absolutely clean but may contain slightly webbed wool.

- Pure white clean, matted wool.
- All stained and unclean wool matted or unmatted. If it is to be stored, both balls or moth crystals in cloth sacks should be placed with the wool in tightly covered containers.

While the above have been the usually acceptable commercial grades of Angora rabbit wool, some grading systems have been simplified to the extent that only 3 grades are considered.

No.1 - Clean wool

No.2 - Clean mats

No.3 - All other wool including soiled wool.

However, wool yield at test shearing in grams in the different classification (ALP, Germany) is as follows:

1. Grade = over 6 cm long, pure, white clean.
2. Grade = upto 6 cm long, pure, white clean. Felt = matted wool.

Table 2 :

	Mean Value	Range
Shearing time (min.)	15	10-20
Other lobour expenditure per Animal/ year (min.)	60	50-90
Production life of Angora Rabbit (years)	4	3-5
Amount of hair –normal haired Rabbit, 2-5 kg of live weight (g)	45	35-50

Table 3 :

Males	Female
140 g 30 points	220 g 30 points
150 g 30.5 points	230 g 30.5 points
160 g 31 points	240 g 31 points
170 g 31.5 points	250 g 31.5 points
180 g 32 points	260 g 32 points
190 g 32.5 points	270 g 32.5 points
200 g 33 points	280 g 33 points
210 g 33.5 points	290 g 33.5 points
220 g 34 points	300 g 34 points
230 g 34.5 points	310 g 34.5 points
240 g 35 points	320 g 35 points
250 g 35.5 points	330 g 35.5 points
260 g 36 points	340 g 36 points
270 g 36.5 points	350 g 36.5 points
280 g 38 points	360 g 37 points
290 g 40 points	370 g 37.5 points
300 g 42 points	380 g 38 points

Table 4 :

	Mean Value	Range
Number of wool hair per cm^2 coat	15000	6000-20,000
Diameter of wool hair-females um	12	10.5-13.0
Diameter of wool hair-Males um	11	9.5-14.0
Diameter of kemps um	28	- -
Percentage of kemps (German Angora) %	2	0.5-5.0
Growth of wool hair mm/day	0.7	- -
Yield %	98	97-99
Wool production/year-females g	1000	800-1750
Wool production/year-Males g	900	600-1450

Table 5 :

Percentage of wool	Males	Females	Males	Females
Over 6 cm long	65	75	6-90	4-89
Upto 6 cm %	27	20	-	-
Pelt %	8	5	-	-
No. of shearings per annum		4	4-5	4-5

After grading, each grade is put in a separate paper bag but not too tightly packed. If the wool is to be stored. It should be stored in a dry place. Moth balls or moth crystals in a small sack before placing the wool in the containers are put to protect the wool from moths.

Some Angora breeders spin the wool on an old fashioned spinning wheel and knit the yarn into garments for home use or for sale. Others sell the wool to organization or individuals who collect large quantities for sale to mills.

Chemical Shearing

Administration of certain chemicals causes the wool to fall out. Chemical shearing has been investigated with sheep. The tropical forage plant Leucaena leucocephala contains an amino acid, mimosine, that causes hair to fall out. Feeding this forage to sheep has been used as a depilation procedure. These treatments have not been studied with Angora rabbits, but would seem to offer a good means or reducing the labour requirements.

Care of Sheared or Clipped Rabbits

The sheared or clipped rabbit will need protection from low temperatures, so ¼ to ½ inch wool should be left on its body. This can be accomplished easily when shearing designed blade is available for this purpose. If the ambient temperature drops to around 35^0 F, the rabbits have to be housed in warm quarters until the wool has grown at least one inch in length.

In regions where the winters are especially severe the programme should be so scheduled that the animals will not be sheared during the coldest months.

Wool Shedding and Wool Block

Wool shedding in Angora rabbit was reported due to the deficiency of protein, crude fibre, copper, zinc, magnesium and iodine.

Wool block is caused by the rabbit licking its coat and swallowing some of fibres. These are indigestible, and they collect in the stomach and form a matted mass. The rabbits fail to eat and become listless, and in a well advanced case the stomach will be firm and hard to touch. As food cannot pass from the stomach into the intestine, the animal gradually dies of starvation within a few days. No effective treatment is possible for advanced cases. If one observes an Angora to be off feed, the stomach should be palpated through the abdominal wall. If a wool block is detected, changing the ration to green feed, root crops, or good quality hay, and massaging the stomach two to three times a day may prove beneficial. Regularly shearing or clipping the Angora and allowing access to a good quality hay and some green feed or root crops will act as a preventative.

Fur Chewing

Fur chewing in rabbits has been associated with a number of factors, improper nutrition, over crowding, high humidity, animal incompatibility, mixing of litters, in adequate ventilation, inadequate feed and feeder space and genetics. Dietary causes are improper amino acid and minerals balance. Pelleted rations should contain at least 15% crude fibre and 17% crude protein. Deficiencies of fibre or protein may result in fur chewing (Cheeke *et.al.*, 1982). Selection of breeder stock should be made from animals that have no history of fur chewing.

Textile Qualities of Angora Wool

New Technologies for processing Angora wool have created new areas of application by means of finer spinning (up to 18 m/g wool). Both demand and price development of Angora wool, therefore, shows an upward trend.

The wool can be blend with other fibres in the manufacture of children's clothing, sport clothes, sport clothes, garment trimmings etc. Used alone, it is usually too light and fluffy, besides, blends create better tensile strength and durability.

It has been observed that 25 % of pure Angora wool in the spun yarn gives the best results with respect to softness, handle strength and luster to the garment. The problem of angora hair shedding off the garment is also not encountered in this proportion of angora wool in blended yarn.

The term Angora is exclusively used in textile terminology to induce that the product is from Angora rabbits only. This is standard by International Standards Organisation (ISO) and is indicated by letters WA (Wool Angora). This is to distinguish it from Mohair of Angora goat WM (Wool Mohair). Ordinary rabbit fur is indicated by letters HK.

The fundamental properties of textile fibre are its length linked with fineness and its suppleness so as form a fine thread.

The Angora wool is the finest with a diameter of 15 u for the duvets; smooth and minimum length is 6 cm.

Angora fibre is made up of keratin which has excellent mechanical properties like elasticity, plasticity, resistance to breakage. Besides, keratin can absorb water upto 33% of its dry weight without appearing wet. The keratin has affinity for different dyes and hence it can be coloured.

Factors Affecting Wool Production

The wool production performance is essentially affected by the following factors:

1. Genetic predisposition (cf.3.2).
2. Energy and protein supply- thioamino acid content of the ration not less than 0.6%.
3. Sex - the wool production of females is approximately 20% higher than that of males; that of castrated males is approximately 10% to 12% higher than that of un castrated bucks, or 8% bellow that of females. The sex-linked differences widen with improving nutrient content in the diet.
4. Intervals between shearing : higher feed consumption with improved thermoregulation results in faster wool growth and shorter intervals between shearing.
5. Age -for each year of life wool production lessens by 10% to 15%.
6. Temperature - at 5°C wool production is 6% higher than at 15° and 22% higher than at 30°C.

Management

For the most part, Angora rabbits are cared for in the same manner as the normal - coated breeds, but there are a few special features and for producing wool that must be taken into consideration. Therefore, raising Angora rabbits is different from raising other fur rabbits.

The bucks and does are kept in individual cages. Woolers (rabbits maintained primarily for wool production), does and castrated bucks can be kept in groups or colonies to save labour. The bucks that are to be used as woolers should be castrated at the time they are weaned. Castration reduces fighting in the herd. However, the wool production is not increased by castration.

The young may be left with the doe until 8 weeks, but only 4 to 6 young per litter should be left with the doe, depending on her capacity for nursing and taking care of them. The doe can be rebred 42 days after kindling.

Because of possible contamination of the wool with hay, dust etc., Angora rabbits should be fed hay. They should receive a complete pelleted diet.

Nutrition

The ration and methods followed for feeding other breeds will meet the requirements for maintenance, development, reproduction and fur production, but as there is a tendency for the average caretaker to over feed Angoras, caution must be exercised to prevent the animals from putting on excess weight. Angoras are not heavy feeders and less than most other rabbits.

Feed and care of the Angora is the same as for other breeds. Because of their long wool, however, Angora must be handled to find out how much flesh they are carrying, the amount of flesh can be determined by running the hand along the backbone. The quality of feed should be reduced or increased to keep the animal in healthy condition.

According to Rougeot and Thebault(1981) the nutrition of Angora rabbit presents some peculiarities because of the difference in its raising when compared to the ordinary rabbit.

Fur is composed primarily of protein, thus a high nutrient feed is appropriate for its best growth.

If the cost advantages of absolute rabbit feed are to be utilized effectively, a feed supplement must be administered to ensure a high level of performance. This high energy feed supplement containing thioamino acid meets the high requirements of the animal after shearing and during growth.

The angora rabbit produces one kg of pure wool with approximately 30 kg of starch equivalent while sheep requires 80 to 90 kg of starch equivalent.

The Angora rabbit should produce per year more than 2 kg of dry proteins. One should count more than one kg for keratin (fur) and also the protein forming hair follicle;

This explains the high quality of nitrogenous matter which is necessary in the ration. High quality Lucerne hay (proteins and vitamin A) complete with oats is preferable. For high wool production 100 g of Lucerne hay and 100 g of oats per day per rabbit is recommended. If good quality lucerene hay is not available or sufficient, a good concentrated diet comprising oilseed cake, dehydrated lucerene, cereals, a supplement of minerals, vitamins and methionine should be fed.

Where granulated feeds are used, at least 160 g to 170 g per day per rabbit is needed containing at least 17% crude protein.

The needs of Angora rabbit follow the cycle of harvest (all the 3 months)and the growth of wool. The needs of the rabbit are more after plucking because it has lost wool. Further, the loss of calories due to heat radiation is more. In the second month, the animal is covered with growth wool but this is the period when the wool grows most rapidly too and so the ration given should be sufficient. In the 3rd month, the needs are reduced because the wool growth is less rapid and also that it is approaching harvest again. Therefore, one should adjust very carefully the daily ration according to these different periods. The feeding schedule should be 190 to 210 g of feed (containing 17%crude protein) per day during the second month, 140 to 150 g per day during the 3rd month. It is equally recommended to fast the rabbits a day in a week and given only straw. So as to permit the stomach become empty and to avoid or diminish the risk due to accumulation of ingested wool if any. The ingested wool form into very hard balls named trichobezoars which obstruct the pylorus, resulting in the death of the animal due to fur block.

Table 6 : Requirements for maintenance and wool production

Performance	Feed (kg)	TDN (kg)
Per kg wool - does	55	35
Per kg wool - bucks	65	41

Comparison between Angora and Sheep Wool

Angora differs from sheep wool in the following aspects:

1) The hair is very fine with a minimum diameter of 11 microns (9-14 um)
2) Low specific weight (1.15-1.18) g/cm^3 compared with 1.33 g/cm^3 for sheep wool and 1.50 gcm^3 for cotton.
3) Heat storage ability approximately twice as high as of sheep wool.
4) High content of pure wool (97%-99%).
5) Eight times more insulative than sheep wool.
6) Four times better wearing quality than sheep wool.
7) Sheared 3-4 times per year as compared to 1-2 times per year in sheep.
8) Corresponding to the body weight, the average raw wool production per annum of the Angora rabbit is about 4 times as high as that of sheep (0.20-0.25 kg wool/kg live weight compared with 0.050-0.07 kg wool/kg live weight for sheep).
9) If the higher content of pure wool is taken in to account, the production advantage of the Angora rabbit is 7-8 times greater.
10) In contrast to the Mohair from Angora goats, the quality of Angora rabbit wool does not deteriorate with advancing year (hair diameter, percentage of guard hairs).
11) However, due to the need for individual housing and time required for shearing the expenditure on work lobour per kilo Angora wool is substantially higher.

CHAPTER 11

Rabbit Diseases

Normal Physiology

Rectal Temperature	:	38.6 to 40.1°C (AV: 39.5°C)
Respiration/Min	:	32 to 60 (AV: 51)
Whole blood volume	:	55.6 to 57.3 ml / kg body wt.
Blood pressure	:	Systolic: 110 (90 -130)
	:	Diastolic: 80 (60 -91)
Blood pH	:	7.35 (7.21 -7.57)
Heart beat /min	:	306 -333 (New born: 220)
Hb (%)	:	8.4 to 16.3 (Av: 11.6%)
RBC 10^6 cmm	:	4.84
WBC 10^6 cmm	:	7.25

Neutrophil %	: 33.5 }	
Eosinophil %	: 0.74 }	
Basophil %	: 2.46 }	But there is a wide variation
Lymphocyte %	: 62.6 }	
Monocyte %	: 0.4 }	

Bacterial Diseases

1. *Pasteurellosis* : (caused by Pasteurella multocida)
   Gram -ve , Non spore forming, short, bipolar rod.

   Other Diseases are:

   i. Snuffles
   ii. Enzotic pneumonia
   iii. Ootitis media
   iv. Genital infection
   v. Abscess
   vi. Conjunctivitis
   vii. Septicaemia

   i) *Snuffles* : Serious mucous/muco purulent nasal discharge, abnormal respiratory sound, sneezing, coughing, reddish nasal mucosal lining.
   Treatment: Penicillin with streptomycin injection,
   Sulpha quinoxaline 225 gm/tone feed.
   Furazolidine 50 gm/tone feed.

   ii) *Enzootic Pneumonia* : Consolidation of lungs, atelectasis (collapsed alveoli) abscess, small nodular focal, hemorrhage with in the lung parenchyma.
   Treatment : Antibiotics.

   iii) *Ootitis Media* : (Torticlos or Rye neck) caused due to mites (Psarcotic or Psoroptic cuniculi) and further by P.multocida.

Treatment : Cleaning by Ascabiol or 0.1 ml Ivermectine S/Cly.

iv) *Gentital Infection :*

Female : Seruos, mucous, mucopurulent vaginal discharge, Uterus filled with suppurative exudates (pyometra), metritis, vaginitis, vulvitis, cervicitis, mastitis,

Male: Orchitis, epididymis, prosthitis , balanitis.

Treatment : Antibiotics

v) *Abscess* : Sub cutaneous swelling, containing creamy exudates surrounded by fibrous capsule depending on duration.

Treatment: opening, dressing, antibiotics.

vi) *Conjunctivitis* : Eye lid swollen, reddish conjunctiva, serous, mucous, mucopurulent exudates, congestion of abdominal and thoracic organs, hemorrhage beneth serous membranes and sub-cutaneously.

Treatment : Eye lotion, eye ointment, antibiotics.

vii) *Septicemia* : Congestion of abdominal and thoracic organs, hemorrhage beneth serous membranes and sub cutaneously.

Treatment: Not possible as they die suddenly, eliminate stress etc.

2. *Tularemia* : Is a normally occurring acute febrile septicemia disease caused by Francisella tularensis. It is grame -ve , non spore forming, aerobic and pleomorphic.

Symptom: Anorexia.

P.M.Findings: Spleen: enlarged (splenomegaly), dark red, may have pin point white foci.

Liver: Congested, numerous white foci.

Lungs: congested, patchy areas of consolidation.

Treatment: Not possible, better eradicate rodents.

3. *Pseudo Tuberculosis :* Is a chronic debilitating disease caused by Yersinia pseudotuberculosis. It is gram -ve, motile, pleomorphic with coccoid to bacillary morphology.

   Symptoms: Diarrhoea, enlargement of mesenteric lymph nodes, emaciation, some times fever.

   P.M.Findings: Enlargement & necrosis in the payers patches of small intestine, caseous necrosis in caecum, enlarge mesenteric lymph nodes / spleen, necrosis of liver / spleen / kidney / vagina / uterus.

4. *Bacillosis :* It is a sporadic condition characterized by necrosis, ulceration and abscessation of the skin and sub cutaneous tissues specially in the region of face, head and neck.

   Caused by: Fusobacterium necrophorum, gram -ve , non spore forming, anaerobic , pleomorphic , usually occurs in long filament, but commonly in coccoid form.

   Symptoms: Swelling and necrosis with abscessation and ulceration involving lips, skin of the face , head, neck, metacarpal and meta tarsal areas and planter surface of feet, open lesions have foul odour.

   Treatment: Sulpha drugs, tetracycline.

5. *Salmonellosis :* Uncommon, characterized by septicemia, rapid death with diarrhea. Caused by Salmonella typhimurium or Salmonella enteritidis, gram -ve, aerobic, non spore forming rods.

   Symptoms: Anorexia, fever,diarrhoea, abortion, muco purulent vaginal discharge.

   P.M.findings: Most of the organs congested, petichial hemorrhage in abdominal and thoracic organs and other serous surface. Necrosis of liver , enlarged spleen.

6. *Tyzzer's Disease :* Characterised by profused diarrhea, dehydration. Caused by Bacillus piliformis. Death occurs with in 12-48 hrs.

P.M.findings: Reddened caecum, colon, petichial hemorrhage in luman of caecum, colon filled with brown watery fecal material, liver contains numerous pin point pale foci.

7. *Listeriosis :* It is a septicemic disease causes sudden death, abortion, caused by Listeria monocytegens, small gram +ve bacteria.

   Symptom: Anorexia, loss of weight or die with out symptom. Some times torticolis, incordination, rolling.

   P.M. Findings: Congestion of visceral and thoracic organs,

   Liver: numerous pin point pale grey foci, Metritis, foetal degeneration if not aborted.

   Treatment: Pencillin / streptomycin.

8. *Tuberculosis* : Rarely occur, characterized by granulomatus inflammation of lungs, digestive tract, kidney, liver, spleen and lymph nodes. Caused by: Mycobacterium tuberculosis , gram -ve , rod shaped, aerobic, acid fast bacterium.

   Symptom: Anorexia, emaciation, may be rise of temperature, occular leisions, enlargement of elbow, stifle, hock joints, posterior paralysis.

9. *Treponematosis :* (COMMONLY called Rabbit syphilis/ vent disease/ or Spirochetosis), Caused by Treponema cunicull, which is a slender spiral shaped bacterium.

   Symptoms: Inflammation of genitals , face and lymph nodes, redness and edema of external genetalia exudation of serum of hemorrhage.

10. *Staphylococcosis :* Very common, caused by Staphylococcus aureus, gram -ve, occour in irregular shape. Characterised by fatal septicemia or suppurative inflammation in nearly any organ or site.

Miscellaneous Bacterial Diseases

i) *Collibacilosis* : Due to Escherechia coli, out break of diarrhoea with high mortality, mucoid enteritis.

ii) Disease associated with proteus vulgaris: it causes sub cutaneous abscess.

iii) Disease associated with Pseudomonas aeruginosa : causes sub cutaneous abscess, septicaemia, pneumonia, diarrhoea.

iv) *Meliodosis*: caused by Mallemyces pseudomallei , nasal and ocular discharge, dyspnea, testes and epididymis frequently contains area of caseous necrosis, and consolidated lungs.

v) *Actinomycosis* : caused by Actinomyces spp. Produce Osteitis and involves mandible , nasal bone, foot, hock joint and lumber vertebrae.

vii) *Hemophilus*: caused by Hemophilus spp-produce firm abscess containing by thigh muscles leading to fibrous capsule. This is gram –ve , non motile , short rod.

viii) *Sylavatic plague*: caused by Yersenia pestis in bush rabbits, hares, desert and mountain cottontails.

ix) *Brucellosis* : rarely occur in rabbits if raised with sheep and goats.

x) *Streptococcus*: associated with acute septicaemia in young rabbits.

Symptoms: anorexia, depression, fever, dyspnea, inconsistent diarrhea.

xi) *Diplococcus*: caused by Diplococcus spp. It causes acute septicaemia, dyspnea, fever.

xii) *Clostridium*: caused by Clostridium perfringens.

Symptom: affecting lactating does, anorexia, diarrhoea, with blood and tympany, ulceration of mucosal surface of small intestine.

xiii) *Erysipelas* : caused by Erysiplothrix rhusiopathiae.
Symptom: nervous symptom, torticolis, abortion, and neo natal mortality.

xiv) *Chalmydia or pittaeosis lymphogranuloma venereum:* causes abortion and pneumonia.

Viral Diseases

1. DNA Virus infection

a) POX Virus infection:

   i) *Myxoma virus*: cause myxomatosis.
   Sym : edema of eye lids, droopy appearance of eye, inflamation and edema around anal, genitalia, oral and nasal orifices, edema around ear, skin tumors.

   ii) *Rabbit fibroma virus*: Symptom- tumors in the muzzle and around eye.

   iii) *Rabbit pox virus:* Symptom – sub cutaneous edema of the mouth and other body opening.

b) *Herpes Virus Infection :*

   i) Herpes virus cunicull:

   ii) *Herpes syvilagus*: not so important in domestic rabbit.

c) *Papo Virus Infection :*

   i) *Rabbit papiloma virus*: Symptom - horney warts on neck, shoulder, eye lids, ear and abdomen.

   ii) *Rabbit oral papiloma virus*: Symptom –appearance of numerous discrete whitish growth on the under side of the tongue.

d) *Adeno Virus Infection :* Adeno virus infection in man is common and respiratory in nature. Hence, possibility of infection of rabbits by man is there.

2. RNA Virus Infection

a) *Piorna virus infection*: causes disease in intestinal tract.

b) *Orthomyxo virus infection*: associated with influenza.

c) Rhabdo virus infection not so important

d) *Paramyxo virus infection*: not so important

e) *Rhabdo virus infection*: affects CNS

f) *Tonga virus infection*: Not so important

3. Diseases Possibly of Viral Origin

i) *Infections vesicular stomatitis* : causes stomatitis, excess salivation, lesion on lips, tongue and hard palate.

Protozoal Diseases

i) *Sporozoa*: (Coccidiosis) like Emeria stiedae/ irresidua/ magna/ perforans, Hepatozon cuniculi.

   Causes: interference in hepatic function, blockage of bile duct, diarrhoea.

ii) *Toxoplasmidia*:

   a. *Toxoplasma*: causes Anorexia, increase respiratory rate, serous or seropurulent ocular gondali and nasal discharge, paralysis of hind quarters.

   b. *Sarcocystis cuniculi* : Produce cyst in the host muscle. The cyst produce strong endotoxin which is lethal to rabbits when administered intra venously.

iii) *Microspordia*: Encephalitozoon cuniculi: symptoms are neurological signs including convulsion, tremors, torticolis, peresis and coma.

iv) *Flagellates and Amoebae*:

   a. *Chilomastix cuniculi* : Non pathogenic, occur in cecum of rabbit.

   b. *Giardia duodenlis*: Non pathogenic, occurs in small intestine.

   c. *Monocercomonas cuniculi*: Non pathogenic, found in serum.

   d. *Trypanosoma* : found in wild rabbits

f. *Trypanosoma sewisi*: like organism: found in wild rabbit.

g. *Entamoeba cuniculi* : Found in cecum, colon. Non pathogenic.

Arthopod Parasites of Rabbits

a) *Mites* :
   1. Psoroptes cuniculi (ear manage)
   2. Sarcoptes scabiei and Notoedres cati: (ear mange)
   3. Cheyletiella spp.(Fur mites)
   4. Listrophorus gibbus (Fur mites)
   5. Trombiculid or Harvest mite: (Red bugs)
   6. Pentastomid mites (Found in internal organs).

b) *Ticks* :
   1. Continental rabbit tick: (Haemaphysalis - leporisplaustris)
   2. Eastern Rabbit tick (Ixodes dentatus)
   3. Hard tick
   4. Soft tick

Flies . Mosquites

1. Suckling louse of rabbit (Haemodipus spp.)
2. Cuterbird flies affect in rabbits.
3. Myiasis flies affecting rabbits.
4. Mosquitoes -affecting rabbits

Helminth Parasites in Rabbits

1. Nematodes

a. *Trichostrongylidae*
   a) Obleiscodes cuniculi
   b) Nematodirus leporis
   c) Longistrita noviberiae

d) Trichostronglyus calcaratus

e) Graphidium strigosum

b. *Metastongylidae*: Protostrongylus boughtoni

c. *Oxyuridae*:

a. Passaluris ambiguous

b. Dermatoxys veligeria

d. *Trichuridae*: Trichuris leporis

e. *Filaridae*: Dirofillaria scapiceps

2. Trematodes

Rabbits appears to be a natural host for few nematodes, but may serve as a reservoir host for several flukes.

a) *Natural host for* : Hasstilesia tricolor

b) *Reservoir host for* :

i. Fasciola hepatica

ii. Dicrocoelium dendriticum

3. Cestodes

Rabbit is a intermediate host for number of types of worms.

a) *Definite host for* :

i. Cittotaenia variabilis

ii. Raillietina salmoni

b) *Intermediate host for:* Taenia pisiforms

Miscellaneous Disease of Rabbits

1. *Pregnancy toxaemia*: Symptoms- Depression, dyspnea, abortion, incorcination, convulsion, coma.

   P.M.findings: Actively secreting mammary glands, large corpora lutea in ovaries, areas of necrosis in the mesentric fat, pale liver/kidney/heart, enlarged pituitary gland.

2. *Mucoid enteritis :*

   Symptoms : Anorexia, subnormal temperature (99-102^0 F) depressed, posture is crouched, rough hair coat, loss of weight, bloated abdomen, gas filled intestine, distended cecum, colon , hard as well as gelatinous mucous feces, squinting of eyes, cold extremities, grinding of teeth , dehydration.

   P.M.Findings:

   Stomach: Distended with fluid and gas and mucosa lined with tenacious mucous.

   Cecum: Impacted with dried contents and gas

   Colon: Distended with gelatinous mucous

   Gall bladder: Distended

3. *Mastitis*:
4. *Sore hock*: Sym. Traumatic ulceration dermatitis of planter surface of meta tarsal region.
5. *Hair balls*: Ingested hair which accumulate and obstruct digestive tract.

Disease of Public Health Importance

1. Bacterial

   a) *Tularemia*: caused by Francisella tularensis transmitted to man by bite of infested arthropods, contact of infected tissues, blood, excreta , eating un cooked meat.

      Mode: Penetrates conjunctiva and skin

      Prevention: Wearing gloves while handling rabbits, use insect replants to prevent bite of arthropods.

   b) *Listeriosis* :caused by Listeria monocytogens

   c) *Brucellosis*:

   d) *Salmonellosis*:

   e) *Pasteurellosis*: caused by Pasteurella multocida by contact

f) *Leptospirosis*:

g) *Pseudotuberculosis*:

h) *Tuberculosis*:

i) *Staphylococcal infection*:

j) *Streptococcal infection* :

k) *Tyzzer's disease* : (by Bacillus piliformis)

l) *Meloodosis*:

m) *Pseudomoniasis.*

2. *Mycotic* : Dermatitis (Ring worm)

3. *Viral* : Encephalitis complex and Myxomatosis

4. *Helminthic* : Common to man and rabbit is Trichostrongylus (round worms)

5. *Arthropods* : Mites, Ticks and Lice.

CHAPTER 12

Disease and Other Health Problems

Health care is an important way of making sure that rabbits stay productive. Every rabbit keeper should know how to recognize and, if possible, prevent and cure illnesses which can affect rabbits.

It is always better to prevent disease incidence than to let animals get ill and then try to treat them. Allowing rabbits to get sick is expensive for several reasons, sick animals stops growing and reproducing so money is lost with this decrease in production, some disease results in the death of rabbits, replacing the dead rabbits is expensive , treatment for the ill rabbits is usually expensive.

Much has already said in this book about preventing sickness and injury. Good feeding and proper housing for example, are as important as anything. They rely entirely on the rabbit keeper to provide everything they need.

Good health should not be based on emerging treatment it should be based on continuing good care:

a. The recognition of health and disease
b. Prevention of disease
c. Resistance to disease
d. What to do if there is an outbreak of disease
   We will also look at the symptoms, causes, prevention and treatment of twelve particular problems:
e. Coccidiosis.
f. Snuffles and cold.
g. Haemorrhagic septicaemia.
h. Pneumonia.
i. Tuberculosis.
j. Salmonellosis.
k. Myxomatosis.
l. Mastitis.
m. Mycosis.
n. Sore feet.
o. Heat stroke.
p. Injuries and abscesses.
q. Internal parasites.
r. External parasites; mange, ear canker,etc.
s. Poisoning.

The Recognition of Health and Disease

A good keeper takes time to look at his animals carefully every day. He is checking to see that they are healthy and are not showing any signs of disease . If anything is not quite right he must recognize it early to prevent it from becoming more serious. What signs should he look for?

Coat

The coat of the animal should be sleek and glossy all over with no dull patches. The shedding of hair (moulting) at certain times of year is quite normal but scurf(loose flecks of skin) should not be present on the coat.

Eyes

The eyes should be bright with no discharge coming from them. The rabbit should appear alert.

Movement

When it moves, the movement should be easy and free. When resting it should be relaxed and breathing evenly and silently at a rate of from 38-65 breaths per minute. The rabbit should not sit huddled in a corner of the hutch , nor should it move stiffly.

Appetite

A normal appetite is a good sign. If a normal animal suddenly goes off feed, the cause should be investigated immediately. It could be an early sign of disease and could be due to a sudden change in diet.

Droppings

These should be normal in amount and appearance . Any signs of diarrhoea should be noted and the cause investigated.

Body

A healthy rabbit is well fleshed in the right areas, the back, loins and shoulders. It should not appear 'bony'.

Weight and Growth

Adult rabbits should maintain the constant body weight, though some breeding females tend to lose weight when they are suckling their young. Young rabbits should grow to reach their mature weight, though it is important to realize that the rate of growth depends on inherited factors and also on the level of feeding. Slow growth may therefore not be due to ill health alone. Loss of weight, or failure to grow, accompanied by a harsh, staring coat is probably caused by disease.

Discharges

There should be no discharges from any part of the body including the eyes, nostrils, mouth, vent, anus, and teats.

Sores and Swellings

The skin should be free from sores and there should be no swellings anywhere.

Temperature and Pulse

The pulse rate of a healthy rabbit is normally 140-150 per minute and the body temperature is 39° C. These are not checked every day of course, but only if a disease is suspected.

Prevention of Disease

The best way to deal with disease is to prevent it occurring by good management. The following measures will help to prevent disease:

Procurement of Parent Stock

Any new rabbits bought in should come from a reputable breeder with a good record of health in his rabbitry.

Quarantine quarters should be established in a place separate from, the other rabbits. They should be used for:

(a) *New arrivals*: These should be kept for four weeks in quarantine to make sure they are free from disease. Rabbits returning to the rabbitry after a period of absence should also spend four weeks in quarantine.

(b) *Rabbits suspected of having a disease*:
If a disease which can be spread from one rabbit to another, such as myxomatosis, is suspected, the affected animals should be isolated in quarantine hutches.

Standards of Housing

Standards of housing should be as high as possible, particularly with regards to the following:

(a) Prevent cold, damp and draughts.
(b) Maintain good ventilation.
(c) Avoid overcrowding.
(d) Keep out vermin (rats, mice, etc.)
(e) Control insects.

Create Hygienic Conditions

By thorough cleaning on daily / weekly basis disinfection of:

(a) Hutches
(b) Floors and walls of buildings
(c) Feeders and drinkers
(d) Other items of equipment.

Nest boxes should be thoroughly cleaned and disinfected between litters. The doe will usually keep the nest box clean whilst the baby rabbits are in it , but if the box is getting dirty, carefully change the bedding. The danger of disturbing the baby rabbits is that the doe will reject them. But if the nest box gets too dirty, the baby rabbits will probably become ill and die anyway. Check the doe carefully; if she is letting her litter get dirty, she may be ill herself.

It is wise to establish a routine for cleaning and to keep to it. Also wood-wire hutches should be cleaned out every day and wire cages whenever necessary.

An outbreak of disease will make additional cleaning and disinfection necessary. Disinfectant should always be used at the correct strength according to the manufacture's instructions.

Good Feeding

Feed the rabbits as well as possible with special regard to the following:

a) Quantity of feed
b) Nutritional quality of feed
c) Clean and fresh food.
d) Regular feeding
e) Avoid sudden changes of food.

Resistance to Disease

All animals have certain amount of built in resistance to disease. If this resistance is lowered in any way the animal becomes more likely to catch a disease. There are many things which may lower the resistance to disease, some of which are due to bad housing and management, such as cold, wet, draughty conditions and poor feeding.

The presence of one disease may also lower the resistance so that the animals get another disease as well. For example snuffles may lead to pneumonia, particularly if the rabbit is cold or damp or poorly fed.

Action to Take on the Outbreak of Disease

If the measure outlined in the Previous section are taken, a serious out break of disease will be unlikely. However disease can occur even in the best managed rabbitries and the rabbit keeper must be prepared to deal with it. Careful observation of his stock every day will enable him to detect signs of possible disease at an early stage. Action may then be as follows:

Isolate

Any animals suspected of having an infectious or contagious disease. This means that they should be kept in cages in a totally separate place from all the other rabbits including wild rabbits. During your daily routine, you should attend to your daily routine , you should attend to healthy rabbits first, then go to your sick rabbits in the quadrant area. If possible wear different clothings sitself, when treating the sick rabbits and leave them hanging up in the quarantine building when you tinish. Put a bucket of disinfectant to treat the sick rabbits and thoroughly wash your hands, arms.

Thoroughly Clean and Disinfect

The hutch and all equipment which has been used for the affected must be thoroughly cleaned and disinfected.

Comfort

Make sure the sick rabbits arc comfortable to give them best chance of recovering from the disease.

Observation: Observe the animals carefully for any symptoms.

Diagnosis

If possible, observe the problem and take appropriate measures. In the case of valuable breeding stock, call the veterinarian or extension agent.

Kindling

The birth of the young rabbits is known as **'Kindling'** If it's a natural kindling, it often takes place at night and morning. It may not require any interference by the keeper. The process usually completes within 7- 30 minutes. This process can be induced with oxytosin injection which enables the proper care of the young ones at the time of their birth. The doe should be left undisturbed. In case she is scared at the time of kindling, she may kill the litter or abandon it.

After delivering a baby rabbit, the mother licks it and allows it to suckle. If all the delivered baby rabbits get a chance to suckle just after birth, the chances of their survival become more. Soon after the mother has nested the baby rabbits, it can be inspected if all the newborn rabbits have suckled or not and can help those who have not.

Good does deliver their young ones in the nest box and are careful about the litter. But sometimes, in certain cases, the doe delivers her babies outside the nest box, on the wire floor. In such cases the rabbits are destined to die of cold. Care has to be taken for their better survival.

References

1. Contera, C. 1991. Systèmes modernes non conventionnels d'élevage du lapin en Espagne. Conférence Duquesne-Purina, 28 November 1991, Chartres-de Bretagne, France.

2. De Lazzer, M.J. & Finzi, A. 1992. Technical and economical efficiency of an unconventional rabbit breeding. Fifth World Rabbit Congress, Vol. A, p. 615-620.

3. Finzi, A. 1992. Rabbit production in developing countries. Fifth World Rabbit Congress, Vol. A, p. 86-94.

4. Finzi, A., Tani, A. & Scappini, A. 1988. The Tunisian non-conventional rabbit breeding systems. Fourth World Rabbit Congress, Vol. 1, p. 345-351.

5. George S. Templeton. Domestic Rabbit Production. The Interstate, Printers and Publishers, Inc, Illinois.

6. Lebas, F. (1980). Les researches sur l' alimentation du lupin : Evolution. all cours de 20 dernieres annees et perspectines d' avenir. Proc. World Rabbit congr., 2nd vol. 2, pp. 1-17.

7. Lang, J. (1981). The nutrition of commercial rabbit. 1. Physiology, digestibility and nutrient requirements, Nutr. Abstr. Rev., **51**, 197-225.

8. National Research Council (NRC). (1977). "*Nutrient Requirements of Rabbits*". National Academy of Sciences, Washington, D.C., USA.

9. Partridge, G. G., Fuller, N. F. and Pullar, J. D. (1983). Energy and nitrogen metabolism of lactating rabbits. Br. J. Nutr., **49**, 507-516.

10. Peter, R. Cheeke. (1987). "*Rabbit Feeding and Nutrition*". Academic press Inc. (London) Ltd., 24-28 Oval Road, London.

11. Ramadhar Singh. (2005). Scientific Production and Management of Goats and Rabbits. Kalyani Publishers, New Delhi.

12. Slade, L. M. and Robinson, D. W. (1970). Nitrogen metabolism in non-ruminant herbivores. 2. Comparative aspects of protein digestion. J. Anim. Sci., **30**, 761-763.

13. Thébault, R.G., Rougeot, J. & Bonnet, M. 1981. Aménagement et équipement du clapier destiné à l'élevage du lapin Angora. Cuniculture, 8: 203-208.

14. Van Soest, P. J. (1982). "*Nutritional Ecology of the Ruminant*". O and B Books, Inc., Corvallis, Oregon.

INDEX

A

B

FSC
www.fsc.org
MIX
Papier | Fördert
gute Waldnutzung
FSC® C083411

Zeitfracht Medien GmbH
Ferdinand-Jühlke-Straße 7
99095 Erfurt, Deutschland
produktsicherheit@kolibri360.de